A Year in the Life of the Yorkshire Shepherdess

Amanda Owen grew up in Huddersfield but was inspired by the James Herriot books to leave her town life behind and head to the countryside. After working as a freelance shepherdess, cow milker and alpaca shearer, she eventually settled down as a farmer's wife with her own flock of sheep at Ravenseat. Happily married with nine children, she wouldn't change a thing about her hectic but rewarding life. She and her family have appeared in ITV's *The Dales* and in Ben Fogle's *New Lives in the Wild*. Voted Yorkshirewoman of the Year by the *Dalesman* magazine, she is also the author of the top-ten bestseller *The Yorkshire Shepherdess*.

Also by Amanda Owen

The Yorkshire Shepherdess

A Year in the Life of the Yorkshire Shepherdess

AMANDA OWEN

PAN BOOKS

First published 2016 by Sidgwick &Jackson

First published in paperback 2017 by Pan Books
an imprint of Pan Macmillan
The Smithson, 6 Briset Street, London EC1M 5NR
EU representative: Macmillan Publishers Ireland Ltd, 1st Floor,
The Liffey Trust Centre, 117-126 Sheriff Street Upper,
Dublin 1, DO1 YC43
Associated companies throughout the world
www.panmacmillan.com

ISBN 978-1-4472-9526-6

27

A CIP catalogue record for this book is available from the British Library.

Typeset by Ellipsis Digital Limited, Glasgow
Printed and bound by CPI Group (UK) Ltd, Croydon, CR0 4YY

Visit **www.panmacmillan.com** to read more about all our books
and to buy them. You will also find features, author interviews and
news of any author events, and you can sign up for e-newsletters
so that you're always first to hear about our new releases.

To my family

Contents

Introduction

'Thoo won't rein sa lang up the'er, mi lass,' said one old boy at the auction leaning over a gate whilst charging his pipe. 'It's as bleak an' as godforsaken spot as thoo could wish for.'

It is now more than twenty years since I first arrived at Ravenseat, and the beauty of the place comes fresh to me every time I climb the moor and look back across this broad sweep of Yorkshire countryside, with the ancient stone farmhouse and its outbuildings below me. It is tough terrain, bleak and unforgiving in winter. But it is grand and inspiring, a place where the seasons and unpredictable weather dictate to us every day of our lives, but where the rewards of life far exceed the difficulties. It is the place where my husband Clive and I rear sheep, cattle and, especially, children.

When I first arrived here, I was twenty-one. I'd been working as a contract shepherdess, living in a tiny cottage in Cumbria with my sheepdogs, a handful of over-indulged pet sheep, a delinquent goat and a couple of horses. I'd found my vocation: I had shunned the urban life that my childhood in Huddersfield had prepared me for and followed my dream to work in the great outdoors, with a dog at my feet and a stick in my hand, out on the hills, shepherding sheep.

When I met and fell in love with Clive I realized I wanted something more: a family. We didn't set out to create a super-size family, but somehow the openness and freedom of this wild, untamed place imprints itself, and filling the farmhouse with the noise and chaos of children seemed the right thing to do. Clive and I work alongside each other on our 2,000-acre farm, caring for our 900 sheep and thirty cattle. Our lifestyle encompasses the whole family: where we go, our children go, travelling for miles in my backpack when they are babies, learning to walk with their hands on the back of a gentle sheepdog, finding pleasure and contentment in the outdoor life that we all lead, whether it is skiing down the snowy fields in the depths of winter when we are snowed in and they cannot go to school, building make-shift dens for themselves in the hay when we crop the meadows in summer, swimming in the dark, peaty tarns, splashing in the icy waterfall behind the farmhouse, inventing games around our docile, long-suffering Shetland pony Little Joe, or riding our horses bareback for miles across the moors.

In my first book I told the story of my early years, how I was seduced by farming life, how I came to Ravenseat to collect a tup (a ram) for the farmer I was working for, and how that was the start of my lifelong love affair with Ravenseat, and the man who farms it, Clive Owen. I wrote about our growing family, and the trials and tribulations that go hand in hand with life on one of the bleakest, most remote hill farms in England.

Our lives are all about routine, but within these seemingly mundane tasks lie many variables: the unpredictability of the animals, the people, and of course the weather. No two days are the same. We relish the contrasts that this life and place bring, from the loneliness and desolation on the moor tops to the con-viviality and warmth of evenings in the hayfields. It is said that in Yorkshire you can experience all four seasons within a day, and the same can be said of emotions. Spirits lifted, hopes dashed, from joy through to the depths of despair, life and death all there in their rawest form. This is what I love, revelling in the challenge of battling the storms, tramping the moors and rearing a family in this inhospitable place. The story of Ravenseat, and of the man, woman and eight children who live here, continues.

1

January

The north wind doth blow, and we shall have snow . . .

Even when the rest of the country is having the mildest of winters, up here it snows. Often it snows for days on end, the snow settling before 'stowering' – being whipped into a maelstrom by the fury of the wind sweeping from the open moors. Then it numbs our faces, stings our eyes and covers any tracks we make almost immediately. When the weather forecasts are reporting a dip in temperatures and wintry showers on high ground, we can be confident that we are in for trouble at Ravenseat.

The arrival of snow brings with it a blanket of whiteness that envelops everything, and the beauty of the changed landscape takes my breath away: even the familiar shapes of the buildings are distorted by the icy cloak, making it all appear so pure and clean. In my first winter at Ravenseat I soon realized why hill farms don't have letterboxes. The snow gets in everywhere, from the gap between the stable doors to the small barn windows. The incursion is not confined to animal quarters, and we stuff rags into keyholes and along sash windows to keep it out.

As long as the weather is calm and settled, the flocks remain on their heafs (the area of the moor which each sheep recognizes as its own terrain, and the only place they are truly happy). We study the forecasts assiduously, always erring on the side of caution, gathering up the sheep and bringing them down from the moor at the first mention of snowfall. We shepherd them down into the more sheltered ground nearer the farmhouse: then we can rest easy, knowing there is no risk of an overnight snowstorm burying them alive on the moor tops. In the worst of times, when even the lower fields are in danger of being happed up with drifting snow, we put the yows (yews) in the sheep pens where they are safe: their constant movement in the confined space means the snow is trampled under their hooves.

It's not hard moving sheep in winter: the rattle of a feed bag triggers a stampede towards the quad bike. The sheep are so addicted to their winter rations that they will follow a feed bag for miles in the hope of a meal of cake, which is a mixed ration of barley, corn, other cereals and vitamin concentrates. The dogs lollop along behind, keeping them in a tight bunch: it would be disastrous for any yow to be left behind in a snowstorm. Only when we have them where we want them do they get the cake, and a feeding frenzy ensues.

When the snow comes, it is not an even covering as the wind has a tendency to change direction, creating deep drifts against the walls – the same places that the sheep naturally move to for shelter. Standing stock-still, heads down and fleeces encrusted with ice, they are sometimes completely covered, disappearing under the peaks, crests and swirls of the crisp white sea. Fortunately it is a rare occurrence for sheep to be buried in a snowdrift, but when it happens we use Bill, Clive's dog, who can sniff out a sheep at twenty paces. We dig them out, but they are not grateful, often stubbornly digging their heels in and refusing to wade through the snow to safety.

One January Clive and I found some tup hoggs blown over in a ghyll next to a wall in the Peggy Breas, one of our fields. When their fleeces are clogged with the driving snow they become heavy, and are easily toppled by the wind. We could see heads, and in some cases legs, sticking out from the ridge of snow. They were all alive and well, but there was no gratitude and we puffed and blowed, sweating under our layers of clothing, as we dragged them by the horns one by one to the gate.

'At least they 'ave 'andles,' I said to Clive.

Every day at Ravenseat starts at 6 a.m., all year round. Clive pulls on his waterproof leggings and wellies and goes outside to start foddering the animals that are in the buildings in the farm-yard. In the winter, when it is still dark at that time, it makes sense to feed and bed up around the yard before daylight, only setting off to check on the sheep when there is enough light to negotiate the precarious routes to the outlying flocks.

The cows, calves and horses are all safely stabled inside over winter, but the sheep stay out, with the exception of a few old or ailing yows. Somehow, every bit of space in the buildings is filled. We wince when visitors occasionally peer over a stable door or into a barn bottom, only to see the very worst of our animals being nursed back to health. It gives a skewed impression – our strongest, best sheep are grazing contentedly out on the hills; the ones inside are the ones who are not thriving.

I get the children out of bed, and breakfast is on the go: everyone eats when they have time, it's not a formal sit-down meal. Porridge, cereal, eggs and toast are laid on: they help themselves. The older children take care of themselves, but I help the younger ones get suited and booted for the day ahead. Raven, who is now fourteen, pulls on her wellies and waterproof leggings over her school uniform, and makes a start on cleaning out and feeding the seven horses. All our children are trained from an early age that waterproof leggings go over wellies: you can always recognize a townie with their waterproofs tucked inside. Hayseeds, sheep cake, rainwater and all sorts of detritus drop into wellies if they are on the inside; plus, you can leave the wellies inside the trousers and pull them on easily the next day. Even little Violet, at five, does it automatically.

Reuben, who is eleven, is in the farmyard from the crack of dawn, feeding calves and helping Raven clean out the horses.

Miles, who is nine, feeds the chickens. He also lights the fire in the black range most mornings. Seven-year-old Edith will, along with Violet, bring logs and sticks from the woodshed: the children soon learn to keep the home fires burning, as the black range heats the water for baths and showers.

'No fire, cold bath,' I say if there are any complaints.

At the same time, I'm filling their lunch boxes. Being part of a big family means none of my children are keisty (picky) eaters. That's not to say that they don't have things they prefer, but with so many people round the table there's always a certain amount of bartering going on.

'Anyone want mi mushrooms?' Miles says, pushing them to the edge of his plate.

'I'll 'ave 'em,' says Reuben. 'But yer'll 'ave to tek mi tomatoes.'

The result of all the trading is that occasionally someone ends up with a plateful of one food. Edith loves carrots, and can eat any amount at one sitting.

'Weeell, she'll allus be able to see in't dark,' says Clive.

Nothing goes to waste. The terriers are on patrol beneath the table, but very little goes their way. Their best chance is to sit under the high chair, and for a time I was convinced that two-year-old Annas was passing all her food to Chalky, who loitered with intent near her place. One morning I watched. I gave her some buttered toast, which she clutched in her pudgy little fingers before dangling it over the side of the chair, where Chalky was poised. The little dog cocked her head, briefly glanced at Annas, then licked the toast. Annas seemed happy with the verdict that the toast was tasty, and reclaimed it for herself.

At the same time as sorting out the lunch boxes, I'm running through the checklist of who's doing what at school that day: do they need their swimming things? Do they need to take their instruments for music lessons? Which of them needs PE kit today?

I check they've put matching socks on: in a perfect world, all socks would be black, and we'd never have a problem. The flake (clothes airer) that hangs from the ceiling above the fire usually holds a supply of lonely socks, and the children are given the job of matching up the odd ones. The flake, also known as 'the sock chandelier', is how I dry the vast amount of washing I do. It means that through the winter months, the flake is constantly in use. I've had many a visitor sitting beside the fire when I've glanced upwards and inwardly cringed at a huge pair of knickers wafting inches above their head, but that was nowhere near as embarrassing as the visiting vicar who had to grapple with a racy bra that got caught in his hair as he sat down for a cup of tea.

'Don't suppose it'll be t'first time 'e's 'ad problems untangling 'imself frae a bra,' Clive said later.

I always do the lunches in the morning: I can't risk it the night before. There are too many predators around here, with Pippen and Chalky top of the list of suspects, though they are not always the guilty parties.

'Summat's been eating my Weetabix,' Edith said the other day.

'Nay, it hasn't . . .' Then I looked and she was quite right, there was a neat hole in the bottom corner of the box.

'I must 'ave torn it when I was bringing it back from t'supermarket,' I said.

I didn't like to admit it, but it had clearly been nibbled by a mouse. I could even see its little teeth marks. In the winter field mice can be a real nuisance, as they come inside for warmth. Old houses have so many cracks and holes in whitewashed plaster walls and skirting boards that it's near-impossible to keep them out.

'I catched six mice t'other day,' I told my friend Elenor, pleased with myself.

'That's nowt – I've got thirty-six,' she said.

I set traps: I don't use poison, mainly because of the danger to the children, but also because other animals could eat the brightly coloured poison granules, or they could eat the corpses of animals who have died because of the poison. The day I gave birth to Sidney I went out to fill the hay racks in the stables and found Chalky lying in a corner, under Josie's manger. Unusually she was stretched out instead of in her normal curled-up sleeping position. I could see she was very ill. Her eyes were sunken, her coat was starey and she was dothering (trembling). I rolled back her lips and looked at her gums: instead of a healthy pink, they were completely white.

I know that rat poison kills by causing internal bleeding, and it was the most probable cause of Chalky's sudden anaemia and dehydration. Quite honestly, I thought she was a goner. Somewhere at the back of my mind I remembered that Vitamin K was supposed to help, and I felt sure I had some in our medicine cupboard. It's given out to new mothers who breastfeed, and I thought that I had some left over from when I'd had Violet.

I rang the vet and described the symptoms, and she agreed that I should give Chalky an oral dose of Vitamin K. She said I had nothing to lose, but that I should then get Chalky down

to the Kirkby Stephen vet practice as quickly as possible. I reckoned I could get her there in half an hour if I put my foot down.

'Aye, I'd say she's taken poison,' Lesley said, looking at
Chalky's gums. 'At a guess I'd say she's been ratting, and maybe
chomped on a poisoned rat. It doesn't take much to kill a small
terrier,' she added as she listened to Chalk's heart.

'No hope?' I asked glumly, biting my bottom lip to stop the
tears welling up.

'Nay, I didn't say that. Here's what we're going to do . . .'

Lesley always has a truck full of dogs of various shapes and
sizes, and it seems one of the prices for living the life of a pampered pooch with the vet is that her dogs are a walking blood
transfusion service: no consent forms to sign, no awkward questions about whether you recently had a tattoo, they're available
to be hooked up at any time. Lesley's lurcher saved Chalky's life.
I had to leave her at the surgery to receive the donated blood
and to be stabilized, and because I gave birth to Sidney (in an
ambulance, near Reeth) later that day, I didn't get to collect her
until two days later, when she was given the all-clear. Chalky was
none the worse for her near brush with death. We reckon she
maybe runs faster now that she's got lurcher blood coursing
through her veins.

At ten past seven the two oldest, Raven and Reuben, get into
the school taxi that pulls into the yard to take them to the
square at Gunnerside, thirteen and a half miles away, where they
wait for the bus to take them to Richmond. It takes nearly two
hours to get from home to school and the same coming back, a
huge chunk out of their day. At first, when they started at
secondary school, they were both very tired, and probably shell-
shocked after leaving the little local school for a comprehensive.
They adapted quickly, so much so that Raven can read a book
while she travels, with no trace of travel-sickness. Reuben reckons

he can do his homework en route, his left-handed scrawl maybe no more untidy than usual. They both seem studious, a surprise considering that Clive and I were half-hearted scholars.

The same taxi arrives back into the farmyard at ten past eight, to take Miles, Edith and Violet to school. The school at Reeth and the school at Gunnerside are run together as a confederation, but they are quite a long way apart. Gunnerside is nearer to us, eleven miles away, while Reeth is seventeen and a half miles away. Because the children split their time between the two schools it's a bit of a logistical nightmare: they are always leaving their shoes, PE kit, sandwich boxes or coats at one school when they should be at the other. Thank goodness Darryl, the taxi driver, has a grip on who goes to which school on which day.

The little ones, four year old Sidney and Annas, are dressed and fed before the taxi goes out of the yard. We wave their brothers and sister off, and then we go out to join Clive in the yard. I have a three-wheeler all-terrain 'running pram' with big bicycle wheels so that I always have somewhere safe to put the little one while I'm busy in the yard, but as soon as I'm on the move the baby goes into the baby back carrier. Once all the mucking out and foddering is done, it's time to load up the quad bike and trailer with hay and feed. Clive and I and the children sometimes go together to the sheep, but more often we go separately, feeding the yows and checking for any that need attention. Time spent observing the flock is never time wasted. A few minutes spent paring an overgrown hoof is well spent: prevention is better than cure.

The weather conditions, and how perilous the journey, dictate whether the children will be in the trailer bouncing around on the hay, or whether they stay down in the farmyard with either me or Clive as we fill bags of cake ready for the next trip.

January is, on paper, a quiet month. If the weather is settled, the cold is not an issue for the flock. There may be little grass for them to eat, but the hay we made in the summer feeds them. We make small hay bales that split easily so that we can spread it out, so that even the shyest of the sheep gets a chance to eat. Also, it's a good way to distribute seeds for next year's crop, reseeding the pastures by entirely natural means, though it doesn't work when the wind picks up and our prized hay can be seen blowing across the moor, balling up and rolling like tumbleweed in a Western, until it eventually disappears from sight. This is wasteful, so then we use round hay feeders. We've got a couple, but we don't like them. The ground around them becomes paddled with mud as the hooves churn it up, and Swaledale sheep have horns which can get stuck when they put their heads through the metal bars into the feeder. Some stand quietly, accepting that they are going nowhere and either eventually extricate themselves or wait for us to come to the rescue; others panic and writhe about, and have been known to contort themselves in such a way that they can suffocate. Trying to free a tight-horned yow stuck in a feeder can be a real battle of wills, the yow pulling backwards with all her might while we try to push her head forwards and down, resulting in nipped fingers. Sometimes the only way to rescue her is to saw her horn off.

We've also got a couple of hay racks on wheels, with mesh grids down the side. The problem with these is that the sheep push their noses through the square gaps in the grid to get the hay, and it ruins the look of their noses. The white marking round the nose becomes more angular and square. The damage cannot be remedied, and the sheep will not sell so well at the breeding sale where looks are paramount.

We start feeding the flock twice a day as soon as the weather gets really rough, though we try to delay this as long as possible, to keep our hay supplies going. But if they can't graze because of the deep snow, they need extra rations.

After all the work we put into our hay crop (you'll read about it in August), I get a bit obsessive about it. I dislike waste at the dinner table, and I dislike waste when feeding the stock. I chase wisps of hay that are blowing down the yard and I once got really annoyed with some resident tups who were overwintering in one of the loose boxes in the yard. Twice a day I took them fresh water, a handful of cake and a few canches of hay which I put in their trough for them to nibble at throughout the day. Every time I looked over the stable door they would be pawing at the hay, just picking at it, and by teatime they would be looking at me with hungry expressions as if to say: 'I'm really ravenous but I've got a lovely comfy bed . . .'

It's not just the fact that it's a waste of good food. Hay is terribly heavy to lift when you have a barn to muck out by hand with a fork. We use straw for bedding, which we buy in, and in the big building we bed the cows on seaves (rushes), which grow in the wetter areas of our hayfields. After our main hay harvest is over, we cut them, dry them and bale them.

When we hear that heavy snow is coming, we keep the children home from school, especially the older ones who have furthest to travel. Lying in bed with the curtains open, we can tell when there's been a significant snowfall. A harder, more unforgiving light is cast, and there's a muffled silence that still perplexes me; why is it so instantly recognizable when Ravenseat is as quiet as can be at the best of times? The peace is soon shattered, as the children rush out there sledging, building igloos, throwing snowballs, and skiing. Before they leave the farmhouse there is a frantic scrabbling in the hat basket for balaclavas, gloves, mittens. They are dressed in layer upon layer of tights, socks, body warmers, overalls, coats, and waterproof mittens on top of gloves.

I bought some second-hand skis on eBay that keep them amused for hours going up and down the slopes. From Violet upwards they are very proficient at downhill racing, but none of them knows how to turn, and the end of each run is the point where they stop or fall over. Muck-spreading can play havoc with the slopes, and hitting a frozen lump of dung at speed results in a few impressive aerial moves. Trudging back up the hill is a chore, so Little Joe the pony or an obliging sheepdog are persuaded to act as ski lifts, with mixed results: the animals don't quite get the hang of where they are supposed to go, and tend to tow the children off in the direction of the stables rather than back up the hill. Sometimes I take pity on the kids, and pull them to the top with the quad bike.

Empty feed bags stuffed with straw make comfortable sledges for the smaller children, though there are inevitably complaints when they hit frozen molehills and are catapulted into the air. The downside of their many layers of clothing usually comes about five minutes after they go outside:

'I wanna wee,' says Sidney, jumping up and down.

'I think tha' Annas 'as pooped,' says Reuben, nodding towards a small figure standing in the farmyard with a look of deep concentration on her cherubic face.

'Great,' I mutter, taking off my gloves.

They may be getting time out of school but they don't get away without doing their homework. We have the internet, and schoolwork is emailed to us. So when the impromptu Winter Olympics is over, and the soggy gloves and hats are steaming on the back of the range, it's down to work. There's no mobile signal at Ravenseat, and we can't have broadband because we are too far from a telephone exchange. The only way we were able to get online was to install a satellite dish. Running a business without the internet is near on impossible, but it wasn't long after the dish was installed that I discovered the social buzz of Twitter, which is infinitely more fun than VAT returns and electronic cattle movements. We were initially dubious as to whether a satellite would survive the rigours of our weather, but it has endured everything, including gale-force winds. Only a direct lightning strike put it out of commission: it was frazzled and needed to be replaced.

At some point, the electricity will go off. It happens every year, usually in winter, when storms bring the lines down. Electricity finally came to the top of Swaledale in the 1960s. It was a very big event and the National Parks authority was opposed to it, because they saw the electricity pylons as a blot on the landscape.

At the time, not all farmers were convinced of the merits of an electricity supply – especially as they had to fork out nearly £300 each to be connected, and that was a lot of money back then.

In the relatively short time I have lived here, electricity has become more and more important in our lives, not always to good purpose. It's ironic that while energy efficiency is such a hot topic, things that once upon a time worked well without electricity now need it. Water used to come out of the taps without it, but now an expensive and complex water purification plant is needed to have a public water tap. Even our septic tank now has an electricity supply which is supposedly so efficient that it is possible to drink the water that drains out of it, but I have no intention of trying. When everything works, it is fine, but when the electricity is off, we're in trouble: you can't even flush the toilet.

High winds are the usual cause of power failure, and it usually happens in the worst of the winter weather, when the children can't get to school. At first it is a novelty, going back to the old ways, and whatever happens we always have a snug farmhouse to retreat to. We will never replace the open fire with anything that relies on electricity.

Outside in the bitter cold, merely touching the gates without gloves is painful, your deadened hands seem to stick to the metal. The water troughs freeze over and we have to break the ice to refill them, carrying buckets of water and bales of hay on our backs. I suffer badly from kins, my finger ends repeatedly splitting and bleeding during these harsh days.

I walk with my head down, to avoid being lashed by the wind that scours down from the hills and whorls around the yard. My thoughts often turn to our forebears, the men and women who farmed this land for centuries past, enduring the same conditions as us but without wellies, waterproofs, and hot showers to revive them, and only smouldering peat fires to warm the draughty farmhouse. People may think we are tough and hard but our life is one of luxury compared to theirs. There were no quad bikes to take them up to the sheep on the moors, they walked miles to market with eggs and cheese to sell. From reading local history books and examining census records, I know that nobody stayed too long at Ravenseat farm in those days: it appears to have been a transitional place, with several smaller farms where we now have one. The families never ventured far, many remained in Swaledale for years, but they generally farmed here as young folk, leaving Ravenseat's exposed and unforgiving land when they could find somewhere just a little bit better.

Much of our energy in winter is spent keeping the animals well fed, but there's a family to sustain as well.

'It's yer belly that keeps yer back up,' they say round here.

With no afternoon teas to do, or breakfasts for guests who stay in our traditional shepherd's hut, I have a little more time to spend in the kitchen, cooking. I fill our big black pot with meat and vegetables and leave it simmering on the traditional black range. Sometimes I bake bread, oatcakes, buns, cakes. I get plenty of practice, because nothing lasts long around here. I make double or triple quantities, reckoning it will last a couple of days, but next time I look in the tins there are just a few remaining crumbs, or perhaps one token lonely bun, as if to say: 'Well, we didn't eat it all.'

Our freezers and our pantry are filled before the winter: we expect to be snowbound at some point, and we need to be self-sufficient. The children love it when I cook on the black range: the smell permeates the whole house. The downside is that in the cold weather they are all ravenously hungry, and they are tortured by the aroma:

'Is dinner ready, Mam?' is a constant refrain.

I prove bread in the ovens of the range, or on the hearth if the ovens are too hot. The little ones watch me testing whether it has risen enough by pushing my finger in and watching the dough bounce back, and when I am not looking they like to have a go. I have found many a loaf with the dents of several little fingers in it.

'Who's been pokin' t'bread?' I ask.

'Weren't me,' is the general chorus.

'Must've been Chalky,' one of them pipes up. A dozing Chalky's ears will flicker in recognition of her name but she'll never move from her comfortable nest under the settle.

The worst time is if the water is off. You go to the bathroom when you wake up, turn the tap and . . . nothing, sometimes just hissing, the sound of air escaping. That is really bad, much worse than the electricity being off. You don't know how much water you use until you have to carry it from the river. Once the water supply to the house is frozen it's impossible to get it back on, as the pipes run under the concrete in the farmyard. We have in the past tried and failed to get it running again, but now we resign ourselves to waiting until the temperature rises. There is a great deal of water-carrying into the house by the bucket-load, heating it in kettles to wash the children as well as we can. Washing long hair is never easy. The water is either too hot or too cold; then you need to heat more water up to rinse the shampoo off.

The pipes to the outdoor water troughs go underground, then up and through the nearest drystone wall, and it is usually just the last bits in the wall that are frozen. There's no electricity supply nearby, so you can't use something sensible like a hair-dryer to gently thaw them. Inevitably, after days of frustration, we resort to the trusty blowtorch, but if you are not very careful you can end up with a burst pipe, so it's better to be patient and stick with the buckets. There is no shortage of water, as Raven-seat has a river close by in every direction. When the troughs are frozen solid we can provide the horses with water by taking them down to the ford; but the old adage about leading horses to water but not being able to make them drink is very true.

Cows drink an enormous amount of water, and it seems like a good idea to take them to the river rather than ferry endless buckets to the troughs. But as soon as they get out of the build-ings they like to stampede up the snowy fields, making them even thirstier. One wintry day, after chasing the galloping, sweat-ing cattle back into the barn, I decided that I'd had enough, and the water troughs were going to be defrosted by fair means or foul. I'd seen a TV programme about Siberia where a fire was lit under a car to get the engine going. It sounded dangerous: fire near to petrol. But I figured fire near to water wasn't such a worrying combination. So I sneaked into the tool shed without Clive noticing: I knew he wouldn't approve of my plan. Matches, some loose straw, and a small bottle into which I'd decanted a little bit of red diesel were all assembled at the other side of the barn wall from the water troughs. I reckoned that I'd soon get a bit of heat into the ground, and the cows would have their drinking water back in no time. I laid my fire carefully, scraping away the snow and the topsoil and getting as near as I dared to the wall. The fire burned hot to start with, fed by wood shavings and anything else I could lay my hands on, like the string and bits of crumpled paper that lurked at the bottom of my pockets. As soon as it had burned for a few minutes, I departed, pleased with myself and sure that Clive would be none the wiser.

A couple of hours later he stormed into the house with a face like thunder.

'Yer know t'water into t'coos?' he said.

'Mmmmm.' I could tell from the look on his face that this was not the moment to boast about my clever scheme.

'I've got serious trouble wi' mi waterworks,' he said.

I resisted the temptation to reply with a smart quip: it was clear from his mood that this was no laughing matter. When I went outside with him I saw a geyser of water spraying high into the air from behind the barn. The whole water supply had to be shut off to investigate and, hopefully, remedy the problem. The frost on the ground was nothing to the frosty reception I got when Clive found the blackened, half-melted alkathene pipe. The weather was too bad to head off to the hardware shop for replacement pipe fittings, so I served my penance carrying far more buckets of water than I'd ever have needed to, if I'd only been patient.

But when it comes to impatience, I'm not alone. I've heard of farmers in the old days, when the pipes were made of copper, putting an electric current through them. It had one of two instant outcomes: thawed pipes, or death . . .

Wood is an essential fuel: we use it to supplement the more expensive coal, which we have delivered. The job of stacking the wood usually falls to Reuben and Miles, with Edith and Violet taking charge of bringing enough into the house to last the day. We are not log snobs: it doesn't matter whether it spits or gives off a good aroma. Forget seasoned hardwood – at Ravenseat we burn anything, including old pallets.

Poor Raven once stood on a piece of pallet with a nail sticking out, which went right through the sole of her welly and into her foot. She hopped around squealing until Clive removed the nail.

''Ow bad was it, Dad?' she shrieked, still hopping.

'It were so far in I didn't knaw whether to use mi fencing pliers to pull it out, or use mi hammer an' just bend it over on't other side. . .' he said jokingly.

Luckily, it hadn't gone deep into her foot; it was a superficial cut and there was no harm done, apart from a spoilt welly.

Every winter we send sheep away from Ravenseat to spare them the worst of the weather. Most of them go at the back end of the year, in November, but we have one small flock that sometimes goes later. We rent a field near Teesside Airport, which has enough grass to sustain about forty-five sheep. We take our thinnest in-lamb shearlings, the ones expecting their first lambs. That's a nice number: just one trailer load, so we don't have to arrange transport on a lorry for them.

It was New Year's Day 2014 when Clive asked me to take them on their winter holiday, and to collect a bull from a neighbouring farmer on the way home. The land at Teesside suits the sheep, with the fields overlooked by a nursing and residential home for elderly people and not used by other livestock. Not only is it good pasture, but the residents take great delight in watching the sheep, counting them, keeping a check on them. There was huge excitement a couple of years ago when one of the yearlings lambed early, while they were still there. The owner of the field, a nearby farmer, casts his eye over them for us now and again, but on the whole they thrive left to their own devices. I take them some mineral lick buckets every few weeks while they are there, but that's all they need.

So on New Year's Day I dropped them off, then decided that, as I was close to a supermarket, I'd pick up some supplies. Our trailer, although in good working order, did have a couple of issues that made it less user-friendly. The handbrake at the front was so stiff that in order to release it I had to jump up and down on it (a peculiar sight) and it had no jockey wheel, which you need in order to attach the trailer to the vehicle towing it. It had met its demise long ago, as a result of our forgetting to lift it when setting off. Consequently it required two people and a fence post to get the trailer on or off: one to lever up the front and the other to back under it or drive away from it. This was a problem that Clive didn't seem to take too seriously.

'I hate this trailer,' I'd say as I reversed up to it yet again in an attempt to get it realigned before setting off. 'It's a nightmare gettin' it hitched.'

'Just don't tek it off then,' was Clive's solution.

This meant that I couldn't just dump it in the field after I'd let the sheep out. No, I had to trail the damned thing with me. Anyway, this New Year I was hungry, and decided to get myself something at the drive-through McDonald's. In hindsight this was not a good idea. There was a fair bit of traffic about considering it was a bank holiday, but I'm used to driving about hauling a trailer, so I didn't worry when I saw the sign about limited headroom. I thought this usually meant no lorries. It was only after I'd shouted my order into the speaker and was edging forward in the queue that I saw that there was an over-hang at the next window, where they hand out the food, and it was far too low for the trailer. There was nothing for it but to reverse out. Now, it seems that a large proportion of the popula-tion of Darlington had also decided to go for a McDonald's, and there was a long queue behind me. I rang Clive, which is my default response in any crisis.

'I's stuck in t'drive-through,' I said. I was harassed and could have done with some kindly, supportive words. All he could do was laugh and say: 'Serves yer reet for not bringing mi a burger. Get a picture.'

Like I was ever going to jump out and start taking photos when there was a crowd of hungry drivers baying for my blood.

What followed was a lot of reversing, plenty of cursing, a few heated exchanges, and a confused member of McDonald's staff, probably handing out the wrong meals to the wrong customers, as I never got as far as cancelling my order.

I eventually extricated myself, and then drove to the super-market. It was nearing closing time as I ran in, and by the time I had hurriedly filled my trolley the light was fading fast. I decided to reconfigure the innards of the trailer in the car park, converting it from the two decks I needed for the sheep to a single deck for the bull. So there I was, scrabbling about in the near-dark, getting sheep muck in my hair, and thinking this was a great way to spend the first day of the year.

'Start as you mean to go on,' I muttered to myself.

I was also thinking about the bull, an animal with a history as far as Clive and I were concerned. I distinctly remembered Clive saying, 'Never again,' the last time we handed Keith the Beef back to his rightful owner.

We have a small herd of Beef Shorthorn cattle, and because

we only have a few, we borrow a bull. Beef Shorthorns are on the rare breeds list, and we like them as they are a native breed. In the same way that Swaledale sheep belong at Ravenseat, so too do the Beef Shorthorns, suited to this landscape, making the most of the rougher grazing and tolerating the weather conditions better than the continental and more modern breeds. They have been around the dales for centuries: a treatise written by an agricultural reformer in 1771 referred to the cattle of Swaledale as 'short horns'.

The bull's official pedigree name is Domino, but it doesn't suit him, apart from the fact he's definitely a black-and-white character. I shouldn't complain about him because in one way, the main way, he did his very best. He impregnated all our six cows, and they gave birth to five healthy heifers and one bull calf, almost doubling the size of our herd in a year. That's a good result. One of the calves was born really late, near Christmas. We had decided its mother was geld, a non-breeder, and because she'd always been rolling in fat we'd never seen her come 'a bullin'', or in season. But clearly she hadn't escaped Keith's attention!

To keep things orderly we name the calves with a different letter of the alphabet each year. The letter for these calves was G, so we named them Gwendoline, Gloria, Grace, Gladys, Gaynor and Gilbert. We raised Gilbert until he was about eighteen months old, then we sold him as a 'store'. That meant he was sold on to another farmer who had the means to fatten him up. Living where we do, with no arable crops and short summers, it would have been a costly exercise for us to fatten him. The heifer calves are kept and have increased the size of our breeding herd.

We had thought about buying our own bull, and Clive and a friend of his went to an open day at the farm of a renowned breeder of pedigree Beef Shorthorns. We were looking for a lighter-coloured roan bull, and there was one on show that Clive really liked.

'Where yer valuing 'im?' he asked the farmer.

'I was looking for around £8,000.'

'Bloody 'ell. We cannae stretch to that, we've only got an 'andful o' cows.'

So it was back to Keith. If my New Year's Day had been rubbish up to that point, it didn't get any better. Keith did not want to come out of his owner's comfortable pen and get into the trailer. When a bull doesn't want to do something . . . He flat-out refused to step off the straw and onto the concrete floor. I felt terribly guilty that on New Year's Day my friend, who was doing us a great favour by loaning us the bull, should now be digging out the bull pen with the muck fork. Eventually Keith succumbed to temptation, taking great strides out of the pen and into the trailer when a bucket of barley was waved in front of him.

Then my day took a turn for the better: our farmer friend handed me a box of chocolates as a belated Christmas present, as well as Keith's passport (all cattle have passports, which is more than I do). Remember, I hadn't managed to get my McDonald's lunch, and I was starving. I'm ashamed to say I munched my way through the chocolates on the long, cold, dark journey home. *All* the chocolates.

Then, feeling guilty (and slightly sick) and not wanting the rest of the family to know I'd eaten the lot, I threw the evidence, the chocolate box and packaging, on the fire when I got in. It was a few months later that I started to look for Keith's passport. I couldn't find it anywhere, and I can only guess I accidentally incinerated it along with the chocolate box . . . Divine retribution: I may have gained a few pounds from gorging on chocolates, but I lost twenty having to buy a replacement passport.

It wasn't long after my trip to collect Keith that Clive agreed the time was right for our trailer to be fettled at the local garage by our loyal mechanic, Metal Mickey. It had nothing to do with the McDonald's fiasco, or with me nearly putting my back out trying to lift the trailer single-handed. The event that prompted the decision to get it repaired came when we were trying to get the trailer on before going to pick up some stray sheep. Clive was doing the levering with the post, and I was reversing. I kept shouting 'Which way?', but Clive is a bit deaf and either didn't hear me, or wasn't listening. I was reading his hand signals in the mirror: left, right, back a bit. Then I saw a few more unorthodox signals: Clive hopping about and putting his hand between his knees. I'd accidentally trapped and squished his finger. He was yelping in pain, and an impressive blood blister later developed. I was not happy that he'd hurt himself, but I was happy with the inevitable conclusion when the next day the trailer went to Mickey's to be fettled.

We don't have a great track record with vehicles, but in our defence, we have to do a fair amount of off-roading. Farm vehicles are real workhorses that suffer far more than the average wear and tear, bumping across cattle grids, potholes, ditches. I was driving to Hawes one day with the children and when I glanced in the rear-view mirror, I noticed that the top part of the mesh door on the back of the pickup had gone. I went cold: I imagined it falling off and taking out the windscreen of a car behind. All the way back home I made the children play 'spot the door' instead of our usual I-spy (something beginning with sh . . . 'Sheep!'). We eventually found it, lying in the middle of the road about a mile from Ravenseat, so it must have dropped off right at the start of the journey.

At least it was the top part of the door that came away on that occasion. Another time I went back to my home town,

Huddersfield, for a family funeral. While I was there I decided to load up with lots of Indian spices and other exotic foods that aren't easily available in Swaledale. I bought a huge sack of onions very cheaply, large bags of turmeric, cumin, coriander and the spicier curry pastes that I really appreciate, coming as I do from the home of some of the best curries in the country. I loaded it all into the back of the pickup, and pulled into the farmyard at Ravenseat late in the evening only to find that the bottom rear door had dropped down somewhere in transit. Not all had been lost, but I had left a trail of onions and spices behind me along the road, and a fragrance that probably puzzled a few people driving the same route.

Losing shopping is one thing; losing animals is a bigger problem. On one occasion I volunteered to transport some Herdwick sheep belonging to our friend Alec. Alec lives relatively nearby, over on Stainmore, and can often be found at Ravenseat, helping out in the sheep pens or doing a bit of DIY around the farmhouse for me. DIY is not Clive's forte, but put him and Alec together and they make a great team, *if* they don't kill each other in the process. Alec is an excellent builder, doing everything perfectly with no margin for error. If he hangs a gate then it will swing properly, the sneck will latch properly and it will never fall off its hinges. Clive, on the other hand, has a more relaxed approach and is a firm believer in six-inch nails, fencing pliers and baler twine, much to the irritation of Alec.

In return for him helping us out we look after his small flock of sheep when his own fields are bare. I was heading for his field at Soulby near Kirkby Stephen, towing the small stock trailer, when I glanced in my mirror and saw a little, rotund grey-and-white sheep disappearing into the distance behind me as I drove past Hollow Mill towards Tailbrigg. The jolting of the trailer had loosened the pins on the trailer's rear door, causing it to drop down and allowing all the sheep to jump to freedom. The only sheep on the moor up there are Swaledales, so these little fellas stood out, and luckily I had Clive's dog Bill with me. Bill really dislikes travelling, so he was very happy to jump out and do the thing he most enjoys: rounding up sheep.

January is the month when we collect a new flock of hens from our friend James's farm. He keeps laying hens, and rings us up every year to tell us when it's time for him to renew his hen stock. The sheds must be cleaned and disinfected and the old birds are given away free to anyone who wants them, so we always collect some, because as our flock gets older they stop laying and eventually die, peacefully, of old age. Nobody keeps tabs on who exactly lays eggs or who doesn't, so there could be some geriatric hens amongst our flock who haven't produced an egg in a very long time, but Miles and Sidney love their chickens. They are in charge of all chicken duties: feeding, cleaning, collecting eggs (with help from Edith and Vi). They were both very excited when Chicken Day arrived. We had to go that day: the lorry was coming at darkening to take the chickens that were left to be processed (I don't like to dwell on this, particularly as I enjoy a chicken curry myself), so this was our last chance.

Miles had spent all Saturday, with some help from Reuben, making a Chicken Rehabilitation Unit for when they arrived. He'd built some nestboxes and filled them with the softest hay he could find, then he'd cut entry holes into some upturned mineral buckets so they could have somewhere dark to hide. He'd hung improvised feeders from the ceilings, and blocked up the window in the barn with hessian sacking to make it warmer: we know the featherless hens we bring back take a while to adjust to life at Ravenseat.

As I looked out of the window on the Sunday morning I groaned inwardly and thought: 'I could really do without this . . .' It was a whiteout. Snow had settled overnight, and there was now a swirling blizzard. It looked set to settle even deeper, and I would have much preferred to stay home. But I didn't want to disappoint the boys, and it would be a full year before we got another chance at the chicken run.

Road conditions meant it would be easier to take the pickup, because if I took the Land Rover I would also need the trailer. Looking on the bright side, I knew I could take the opportunity to go to a supermarket to collect some shopping: it was looking as if we would be snowed in for a few days, and I needed to stock up on supplies. Miles and Sidney came with me, even though I told them we would be going a long way round through Sedbergh and across the M6. They were not going to be persuaded that it would be warmer and more comfortable for them to stay by the fire in the farmhouse: they wanted to select the new hens.

We were only a couple of miles down the road when I heard the first 'Are we nearly there yet?' I stayed patient: it was going to be a long day.

The council usually grits the main roads, but we were on smaller quieter country lanes and it was very early on a Sunday morning so there was nobody else daft enough to have ventured out yet. I had thrown a shovel into the back of the pickup, because I knew from experience that scooping grit from the roadside with bare hands is no fun. The abrasive grit, and the high proportion of salt mixed into it, left my hands stinging and sore for days. Sheep love the salt, and they lick their way through the piles left at the roadside. Salt helps with Vitamin D deficiency in the winter, and we give our sheep salt along with their vitamins, as they aren't anywhere near a road for free supplies courtesy of the council. We buy big lumps of Himalayan rock salt for them to lick.

The previous year, I'd done the chicken run the wrong way round: I'd collected the chickens before I went to the supermarket. The chickens were all trying to escape into the supermarket car park while I was loading in the shopping. So this time I was sensible: shopping first. I go to supermarkets so infrequently that when I get the chance, I make the most of it and fill my trolley. I managed to stow quite a few bags of shopping in the front cab of the pickup, then tied the top of all the other bags carefully, before putting them in the back: I didn't want the chickens pecking at the contents.

Meanwhile Miles and Sidney were sitting in the pickup, temporarily distracted by the warm pies I'd bought them at the deli.

'Not chicken, are they?' Miles asked as he nibbled at the pastry.

We were soon at the chicken unit. Clive had given us strict instructions to get no more than a dozen, remembering the time I set off for twenty and came back with a hundred; but when you see them, and you know what their fate will be if you don't take them, it's hard not to let your heart rule your head. After discussing it with the boys, we took thirty. After all, we've got room for them all in our barns, and they are amazingly productive. I have a whole range of tried and trusted recipes for when we have an egg glut: chocolate mousse, custards, meringues, brûlées, omelettes and French toast.

Miles said: 'Are we going back the same way, Mam?'

I didn't fancy spending any longer on the road than I had to, but to go the other, more direct route meant crossing the boundary between Cumbria and Yorkshire beyond Kirkby, on a notoriously difficult road that is often blocked with snow.

'We'll see. Let's drive up to t'bottom o' Tailbrigg and tek a look,' I said. The warning sign 'Road Closed Due to Snow' had been put out, and there were wheel marks where other vehicles had turned round. *Shall we risk it? Or shall we do the forty-mile detour?* I thought, keeping my fears about the treacherous journey ahead of us to myself.

A car drove up behind us and I was hoping he'd overtake and go up, with me following in his wake. But he took one look at the hill and turned round. I did what I usually do when in doubt: I rang Clive, knowing that this was the last place I would get a signal. Edith answered the phone, but Clive rang back.

'I'm either gonna to be home in fifteen minutes, or in a few hours,' I said.

'Mek sure thoo's in four-wheel drive, back up to t'snow pole an' ger a good run at it, give it all she's got. Don't waffle,' said Clive.

I was tense and nervous, not least because another car had driven up behind us, and the driver seemed to be waiting to see if I was going to tackle the hill. Sidney was oblivious, fast asleep, but I noticed Miles was gripping the sides of the seat. I don't usually mind driving through snow: we do it all the time. But I remember once in another pickup we had, when I'd run out of steam when I reached the top of the hill, the wheels had started to spin and then I knew the horrible feeling of sliding backwards, slowly picking up speed as I struggled to stay in the middle of the road, with the wheels locked, completely out of control, praying that I didn't go over the edge where the crash barrier ended. I was lucky to emerge unscathed that time.

'Right, let's go,' I said, with fake bravado. Miles gave a weak smile, and held on tighter. The chickens in the back were clucking away for all they were worth.

'Nowt to worry about here, Miley,' I said, trying to convince myself as well as him. And there wasn't anything to worry about: we didn't break stroke, climbing the steep hill with snow banked on either side, then negotiating the treacherous road that divides the open expanse of moor. We felt like pioneers, cutting new tracks through virgin snow.

It was all worth it. This batch of chickens was amazing. They laid thirteen eggs on the first day, and although we expected them to go off laying until they got settled and feathered up, they never did.

One afternoon, there'd been a thaw and it had been raining all day. The ground was sodden, the rain chilling, the beck rising. I was on my way to feed the cows and I thought, *I'll feed the chickens so that Miles doesn't have to come out in the cold and rain when he gets home from school.*

He wasn't grateful. He cried. He cried because I had done his job. *Right*, I thought, *never again*. So now I just leave him to get on with it, and I've never had to remind him of his hen duties because he is devoted to them. He constantly looks around the kitchen for scraps for them. One day I was given a few loaves of out-of-date stale bread by a shop in Hawes.

'You can tek the children to feed the ducks,' said Jackie, the shop owner.

'Nah, it'll go to t'chickens,' I said. 'You get summat in return for that.'

Another customer, who hadn't heard our exchange, turned to me and said scathingly: 'Ooooh, fancy saving yerself a few quid by feeding them poor kids on mouldy bread.'

I was furious. It was rude and unfair. But I refused to descend to her level, and managed to bite my lip and say nothing, even though my face probably betrayed my feelings. Jackie's face was also a picture. I muttered about it all the way home, but soon forgot the affront when I saw how happy Miles was with the bonanza for his hens. He is a very self-contained child, happy in his own company. His face usually looks serious, but when a smile breaks across it, it is like the sun coming out. You have to work for his smiles, but they are well worth it.

We let the new hens out of the barn as soon as there is some decent weather. They look terrible for weeks: they have little quills of regrowth which eventually become feathers, and pale combs and wattles. But within a few months they are indistinguishable from the others. Violet loves the egg collecting: the other day, when Clive was in charge of breakfast, she took three eggs to him with the request, 'Boiled eggs, please.' Clive doesn't do precision cooking: I think they were all hard as bullets.

Every four years we have to have our cattle TB tested. We're in a safe area, but we still have to be tested or there are movement restrictions on the cows. The vet lets us know when we are due for testing, which in our case happened last January – the depths of winter. The vet turned up and did the first round of tests, injecting them and snipping off samples of their hair. The second tests have to be done seventy-two hours later: not sixty or eighty, but exactly seventy-two. By that time, a blizzard was engulfing Ravenseat: snow was thick on the ground, and more was falling. We knew that if the vet didn't get to us at the right time, we had problems, as the tests would be invalid and we would not be TB cleared. Luckily, Clive's son Robert, who farms down at Kirkby Stephen, was able to pick up the vet on a tractor and get her up here. It must have been a long, slow journey, but at least we complied with the regulations, and we were declared TB free.

Robert used to run the farm, Sandwath, for us, but he's now taken the tenancy on as his own. We'd wanted Sandwath as well as Ravenseat as it would give us access to more fertile land and a place to overwinter some of our sheep. We found that running two farms is great when everything is in the right place, but we seemed to spend too much time trundling stuff between the two locations. Of course, whenever there's an emergency, or during particularly busy times, we help him out and he helps us out, too.

Our finest breeding yows live out their days at Ravenseat, kept together in a small flock we affectionately call 'the crusties'. They are old dears who have done their absolute best for us over the years, but are now shadows of their former selves. When I looked over the stable door at them last winter it was like peering into an old folks' home: there were a dozen of them, some missing a horn or two, all looking grizzled and ancient. One or two of them hadn't even been put to the tup: I genuinely didn't think they were up to carrying another lamb.

We let them live in the better fields, with the richer grass nearer the farm. They can't go up on the moor any more, and in winter are housed in a barn or stable during the roughest weather. I took a picture of one of our old crusties sheltering with her back to a wall in a snowstorm, and posted it on Twitter. I'd put her and her friend, another oldie who has since died, in the garth by the shepherd's hut. They were eating that much hay and cake I'm surprised they didn't keel over: there's a risk of killing with kindness. It was one of those days when the snow was squalling down, and she had her old grey face turned towards me. It was a nice shot, I liked it, it was atmospheric: very grey yow, grey wall, grey sky. A lady artist contacted me and asked if she could use my photo as a basis for a pen and ink drawing, and her picture ended up in a gallery and then on some Christmas cards.

Clive wasn't impressed.

'Aye, that's right, tek pictures of t'worst sheep on t'farm, why don't you?'

The artistic merit of my photos doesn't come into it: he just wants me to show the sheep off at their best. When a television crew came up to interview us, they asked if there were any sheep inside that they could film.

Clive said, 'No, they're all in t'fields.'

He'd shut the top of the stable door to hide the crusties, and the poor old dears had to stand in the dark until the cameras had gone. I say to him, 'Clive, other farmers an' most people know that not every sheep in the flock is going to look perfect.'

Somehow the sheep with the limp will always end up at the front, holding up its dithering foot, and the one with the mucky arse will always be showing its back end to anyone looking on.

'It'll look fake if we only show t'best looking sheep in pictures,' I say. But he's having none of it.

We've got one yow with a completely sooty face, without the white ring round the eyes that a good Swaledale should have, and whenever anyone takes a picture up here she somehow gets into it. She's like one of those minor celebrities always hanging around to get themselves in a paparazzi picture. Clive goes mad. 'Get that sooty-faced bugger out of t'picture,' he says. It's no good telling him that only another Swaledale breeder would recognize her imperfections. I took a cracking picture of the sheepdogs working the sheep not long ago, but there she was, right at the front, and Clive was not happy. 'I hate that sheep,' he said.

Clive's sheep are like top models: they can be as bad-tempered and as nasty as you like, but everything can be forgiven if they look good.

I was recently asked if I would be happy to pose (fully clothed, I must add) for a professional portrait photographer. As usual, what started out as a simple request became more complicated as it went on.

'Can you sit on a hay bale?' he said. 'I'd like Bill, the sheepdog, in the shot,' he continued as he fiddled with his camera. 'And I'd like you to be surrounded by sheep.'

Not easy, when you have a sheepdog by your side.

'They'll come if I rattle a feed bag,' I said, beginning to tire of the whole enterprise.

'Good idea . . . mmm . . . but you'll have to hide the feed bag, we can't have product endorsements. Maybe you could sit on it?'

'Wouldn't that mek me look like I've got an incontinence problem, sitting on a plastic bag?'

Finally, after an age of him clicking away, the sheep and the dog getting more and more bored and me wishing I was somewhere else, the photographer was satisfied, and left promising to return a week later to let us see the results.

He duly arrived, and we spread the pictures across the kitchen table. Sure enough, there was the sooty yow peeping over my shoulder. That one was firmly rejected by Clive. He didn't care how glamorous I looked, how perfect the backdrop was, or how proud and masterful Bill appeared: any picture with *that* yow on was consigned to the 'out' pile. He nodded happily at the photos of me looking uncomfortable, sitting at a contorted angle to hide the plastic bag, with a drip on the end of my nose. It was all about good-looking, classy sheep. Not a good-looking, classy wife. Which just confirmed where I stand in the natural order of things at Ravenseat.

2

February

'I'm gonna get them two sneakin' buggers back here if it's t'last thing I do.'

Clive was eating his breakfast one chilly morning in February when he made this announcement. The problem had clearly been on his mind.

The weather is bitter up here in February, but however grim and cold it is, the exciting element of the month for the children – and for me and Clive, too – is the big day when Adrian, the ultrasound scanning man, comes to scan the sheep and tell us whether they are in lamb and how many lambs they are carrying. Naturally, we need to round them all up for scanning, which is normally not too difficult, as every day we visit them with hay and cake.

But last February we had a couple of shearlings (yows in their second year) who decided to live in Boggle Hole, the ghyll with sides so steep it is almost impossible to get into . . . unless you are a pair of stubborn sheep. They came out of the ravine every day and joined the rest of the flock for food, but always retreated back down there. They more or less went feral, and were very wary of us, disappearing the minute we appeared. We needed to bring them in for scanning, but all attempts to get them out had failed. However, Clive was on the case, and a plan was formulated. I would put out a small amount of hay near to the ghyll then move away. When the sheep were sure that the coast was clear and they turned their woolly backs to eat, Clive and his dog Bill would sneak around behind them, cutting off their retreat.

It started well: the pair of sheep fell for it, and were soon hoovering up their breakfast. But Bill, who is as fine a sheepdog as there ever was, diligent and faithful, has one small failing. Sheep aren't the only thing he likes to chase. He loves rabbiting, and while Clive was stalking the two wayward sheep, Bill headed off in hot pursuit of Bugs Bunny.

Clive gave Bill his right-hand whistle, a sharp, piercing sound that should have sent him racing in an arc around the sheep to cut off their escape route. Hearing the whistle, the sheep raised their heads, twigged that a trap was being set, and, true to form, set off at breakneck speed back to their lair. It was only then that Clive realized Bill was missing.

'Bloody hell,' he shouted, waving his arms manically to try to turn the sheep, which were heading towards him at full speed. Hearing the commotion, Bill appeared on the rocks above. He's a dog that can read a situation, and he immediately came to Clive's aid. But the sheep had seen their chance and were stopping for nothing. One of them bowled right over the top of Bill, knocking him off his feet and sending him down the rocky scree and out of sight. All we could hear was the sound of rocks falling, then a yelp from Bill, then silence.

'That'll do, Bill,' Clive shouted.

Nothing, just silence. Clambering down as far as he dared, Clive saw the dog laid out on a ledge, not moving. He looked dead. The sheep were forgotten: all that mattered was rescuing Bill. Fortunately, Clive had his stick with him, and by lying down and stretching out, he could hook it through Bill's collar and haul him back up.

There was no mark on Bill, no blood, but his tongue lolled out of his mouth and his eyes were glazed. He was clearly concussed. He's a big dog, and it took a bit of effort from both of us to lift him onto the quad bike and get him back home.

'I'm gonna lie him by t'fire,' Clive said. 'He's in shock. I thought 'e was a goner.'

Pippen and Chalky, the terriers, were evicted from their favourite place to make way for the patient. As he started to come round, I swear I saw a worried look on Bill's face. The *only* time a sheepdog is brought in to lie on the fireside rug is when his days are numbered, and he's on his way to gather the sheep from the Elysian Fields. Bill wasn't having that: the next time we went to check on him he was up on his feet and considering cocking his leg on the fender. He was very happy to be leaving the house.

We did eventually catch up with the two wayward sheep, but only after a show of dog power when Clive enlisted help from his pal Alec and his dogs too.

Adrian, the expert scanner, is in great demand at certain times of the year, which means his days are planned like a military campaign. Not only does he scan sheep, he also contracts round bales, and farms in his own right. When it comes to multitasking and organizational skills, he is second to none. If your allotted time for scanning is three in the afternoon, then heaven help you if your sheep are not penned and the extension cable not at the ready, for efficiency is Adrian's middle name. It is a good idea to have a supply of strong coffee at the ready, and should you want to know anything about any farmer within a hundred-mile radius then you'll likely find out, as Adrian will have scanned their flock.

The whole success of Ravenseat revolves around breeding pedigree sheep. It can be a chaotic day: Adrian gets paid per sheep that he scans, and he doesn't want to waste time hanging about while we sort out the yows. The day before he comes, we start to gather up the yows from the more remote heafs and bring them into the fields near the farm. We don't feed them: full tummies obscure Adrian's view of the unborn lambs.

Every yow will go through the race, a narrow channel that prevents the sheep from turning around, which leads them into the scanning crate. Then the children play 'guess how many lambs' games with one another. The sheep are marked according to their status: no mark means they are expecting one lamb, then there is a coloured mark in the middle of the back for those expecting twins (the colour changes from year to year) and a different coloured mark for triplets. Thankfully, we don't get many triplets: we go for quality, not quantity. Then there is another mark, the dreaded red stripe, for the sheep that are geld: not pregnant.

Not so long ago, Adrian would study the screen as he rolled the scanner under each yow's belly and call out 'Two!', 'One!', 'None!', and the older children would put the mark on the sheep with a spray can. Unfortunately for them, they no longer have this vital role to play, as Adrian has invented a foolproof system that does it automatically. I think he got fed up with mistakes being made, with Clive going 'Eh?' or the children not concentrating and putting on the wrong mark. He decided to remove the weakest link.

Incidentally, I get the same old joke every time: 'Shall I put the scanner over you?'

Anyway, despite not having to spray the marks, the children still enjoy the process, watching for their particular sheep to be scanned. Raven kens a lot of the sheep by now, and there's a couple she always watches for. One is known as Raven's Listeria Lamb, or Roly. Roly's mother, who was close to lambing, was really sick with listeria, and Raven was looking after her in the stable, feeding her cake and treacle and keeping her on her feet. As long as a poorly sheep can stay on their feet, there's a chance of recovery. Being upright keeps their stomach and everything working: once they're down they don't last long. Each day Raven would feed her from a scoop and guide her to the water trough. Listeria affects the brain and balance, and the damage done is almost always irreversible, so we were not surprised when one morning Raven found her collapsed and couldn't get her up. Between us we shuffled her into as comfortable a position as we could, propping her against a hay bale, but it was clear her days were numbered. She wasn't in distress, so we left her there, quiet.

That night, when Rav went out to her again, she'd lambed. It was a beautiful gimmer (female) lamb, perfectly healthy. But her mother didn't make it: she died giving Raven the present of a lovely lamb, a thank you for all her care. Now the lamb is a yow, and every year Raven wants to know what she is going to have.

Miles has his own flock of fourteen Texel/Swaledale cross sheep, Texdales. We often put out a Texel tup with our Swaledale yows at the end of tupping time, to 'jack up', tupping any yow who is not in lamb. A late-born Texel cross lamb is more valuable than a late-born Swaledale. We can sell a smaller Texdale lamb at the auction as a fat lamb and get a better price than a smaller Swaledale lamb ever would. We kept the gimmer lambs for Miles as he wanted something different from our main flock, that he could distinguish as his. They're very hardy, good sheep, and they look great. He had to wait eighteen months until his little flock of gimmer lambs were big enough to be put to the tup. We bought him a Texel tup of his own for £200 at the Luke

Fair sale in Kirkby Stephen. His face was an absolute picture when Clive came back with the tup in the pickup.

'Giz a lift, Miley,' Clive shouted. The tup weighed a ton, and wasn't happy about being manhandled out of the pickup. 'Watch out he don't catch his 'nads on t'backdoor, he's gonna be needing 'em.'

Farm children get to know how nature works from a very early age.

When scanning day arrived Miles, who is very quiet normally, was clearly excited. His yows were scanning for twins one after another, and even Adrian the scanner said to him, 'Miles, if all fourteen scan for twins I'll give you all the money I've made today.'

Fortunately for Adrian the final total was ten twins and four singles, so he didn't have to part with any money, but I'm sure he would have calculated the odds very carefully before making such a wager. Miles didn't mind: he had his lambs to look forward to.

Looking after sheep is a great training for a child: it teaches patience, because nothing happens quickly. There are no instant rewards, like there are in computer games: it's the slow march of the seasons, just as it always has been.

Miles only has one aim: to be a farmer. As far as he's concerned, school is a necessary evil he has to endure until the day he can be free to work full-time on the farm. He's happy enough at school, has plenty of friends, but all he wants is to be here farming. At teatime when we're all gathered round the table eating, I will ask what everyone has been doing at school.

'Geography,' says Raven, rolling her eyes. 'I can't stand geography.'

'Maths,' says Reuben. 'Fractions.'

'Reading,' says Edith. 'I've nearly finished my book.'

'I think Aygill are scalin' t'muck,' says Miles, referring to a farm that he passes on the way to school. 'An' Ronnie was catchin' moudies (moles) this morning. I see'd 'im outta t'window.' He clearly spends a lot of time staring wistfully out of the classroom window.

Outside of sheep and chickens, his big love is fossils and dinosaurs. He goes off fossil-hunting along the river, sometimes with Reuben. Sidney can be found tagging along behind. Sidney loves his big brothers: all he wants in life is to be with them, doing whatever it is they are doing.

Edith is doing very well at school: it seems almost every other day she brings home a certificate for her class work, her behaviour, her spelling, something. 'Edith, you are on fire!' one of them says. She's very proud of them – unlike Miles, who also gets certificates, usually cryptically worded: 'Miles, for being reliable.' He produces them, crumpled, from the bottom of his schoolbag, days later. Edith is very outgoing, very friendly, and has a permanent beaming smile on her face. Violet only started school this year, but she was very keen to go: she and Edith are the best of friends. Vi is a real tomboy: the best climber, kicker, jumper of the lot. She's naturally athletic, always doing handstands, somersaults or wrestling with Sidney.

It was in February a year ago that the headmistress of the school came to see me at the farm. Getting to Parents' Evening was a logistical nightmare for me: appointments were spread out, with long gaps between them, at the two different school sites. So the easiest solution was for Mrs Johnstone to visit me instead. I reasoned it would remind her how remote we are, and perhaps make the school more aware of the problems we have with the weather and the school run.

It was not her first visit here. When she took over the headship, the retiring headmaster brought her up here to meet us. It was an interesting occasion, because our old friend Alec was also paying us a visit: he drops in whenever he likes. He took it upon himself, as only Alec can, to treat these two teachers to his theories of what is wrong with young people today. His remedy seemed to involve acquiring a large cane and beating some sense into them.

'Di'n't do me any 'arm,' he said.

'That's a matter of opinion,' said Clive, as the headteachers tried diplomatically to explain that this isn't the way we do it today.

Anyway, when Mrs Johnstone was due to come again, I decided to have everything just so. I primed Violet, who was not yet at school then but was due to start the following September, to show some of her writing and demonstrate her counting. I had blacked the range, cleaned the house (well, some of it – the visible bits anyway), baked fresh cakes. All would be well. For one day only I was going to be a Domestic Goddess with a Child Prodigy.

But this is a farm, and things never go to plan.

On the day of the visit we had an early start, as Clive went off to Carlisle auction with his friend Mark to sell cattle. I had a couple of lots of sheep to feed, and the usual bullocking up around the yard with Vi, Sidney and Annas helping me. We worked our way through the long list of jobs until there was only one remaining, and that was to muck out the stable of our veteran horse Meg. When I got to her stable I could immediately see she was ill. She was standing with her head bowed, facing the stable wall. There was a green, frothy mucus oozing from her nose, and every so often she'd give a deep, rumbling cough. Her illness had come on so quickly: only the previous evening she had been out until darkening while I filled the hay racks and nets, and there had been no sign she was unwell. My first thought was pneumonia, so I gave her a shot of penicillin and then rang the vet.

'Hello, it's Amanda at Ravenseat. I'm wonderin' if there's any chance of gettin' a veterinary up 'ere?'

I explained the symptoms: snot, coughing and gasping. As I reeled them off I began to question my initial diagnosis, because pneumonia does not normally strike quite so rapidly.

'Actually I'm thinking it might be choke,' I said.

'Someone will be with you within the hour,' the receptionist said.

Next I rang Clive. He was at the auction and I spoke against the background noise and general melee. I wanted to know if there had been any problems the day before, whether the horses had got into the feed store or if anything else had happened that I didn't know about. I got the standard reply: 'Bloody 'osses, nowt but trouble.'

An hour is a long time to wait for medical help. As everyone knows, googling takes you into a minefield of possibilities, so I resisted the temptation to go online and instead resorted to my standby book, *Modern Practical Farriery* (first published in 1875, but horses haven't changed that much since then). The symptoms began to look more and more like choke, and according to the book there was a strong chance the blockage in her throat would dislodge itself. I really hoped so, because I'm in charge of soaking her sugar beet feed. Sugar beet pellets are an excellent food for some animals, like pigs, cows, sheep; but horses need to have them pre-soaked, otherwise they swell up inside the stomach and can kill.

After what felt like forever, the vet arrived – a newly qualified lass who I hadn't met before.

I said: 'I'm thinking it's choke . . .'

She said: 'It could be pneumonia . . .'

She checked Meg over, taking her temperature, palpating her neck and throat, listening to her stomach. She couldn't be 100 per cent sure, but choke was now looking the most likely diagnosis.

'I think we need to get a stomach tube down her,' she said, 'but I haven't got one with me.'

'I've got a calf tube,' I said. 'Mebbe it'll let us see whether we're reet.'

The idea would be to stick it up Meg's nose, down into her oesophagus and through to her stomach, and when it could go no further because of the blockage, pour warm water into the pipe in an attempt to soften the offending lump. Unfortunately the calf tube snapped as the vet tried to bend it in her hands (with calves and cows you can go over the back of the tongue and straight down the throat, but with horses it has to be up the nose, and the tube needs to be more pliable).

She went to the phone and called for another, older, colleague to join her, bringing a specialist longer equine tube. Eventually we were all set: Meg was brought out into the yard and given a sedative to keep her quiet while we did the unpleasant job of getting the tube into her nasal canal. It is not a pretty sight: the tube scraped her mucous membranes causing a slight nosebleed. A funnel was put on the end of the tube, but we had to be sure it had gone down past her windpipe before we put water down: if water got into her lungs it would mean an instant, horrible death. It was very tense.

The pipe was two-thirds of the way down when it refused to go any further, which meant it had hit the blockage. The funnel was held aloft, and jug after jug of warm water was tipped in. What came back up were little pieces of sugar beet and hay, stinking terribly from fermenting in Meg's oesophagus. With each wash out she would shake her head and sneeze violently, distributing bloody snot and saliva on all around. It was difficult to keep the tube in place without getting completely covered.

We now knew for certain what the problem was, but it wasn't about to dislodge easily. Greedy old Meg must have carried on eating and eating, and the food just backed up with nowhere to go. It was clear she would need washing out more than once, and the vet would need to return. The risk was that Meg would be weakened by hunger as she could have no food, just sips of water, and there was a very real chance that she might not survive.

In the midst of all this, I had forgotten about Mrs Johnstone until I heard a car pull up in the farmyard. Moments later she appeared from around the corner. I winced: instead of being clean and dressed to impress, I was covered in horse snot. The children, too, who were watching the procedure from a safe distance with great interest, were splattered with it. I stood still, holding Meg's lead rope, as she snuffled and coughed. There was a pervasive stink from the food she had brought up. The vets were just packing up to go.

'Oh, erm, Mrs Johnstone, how are you?' I said. 'We've 'ad a bad mornin'. Would ta be alreet to just give us a minute?'

I needed to get Meg back into a stable, but I didn't want to put her in her original one: I wanted her in a bigger one, closer to the house with a bigger door (for practical reasons. There have been times when an animal has died and we've been faced with dismantling a door frame or dismembering the corpse). Clive's old tups were going to have to give up their home for her.

We're always playing musical animals in winter, moving things around, but with just the smaller children to help me it was going to be tricky: Clive was going to be pretty annoyed that I'd evicted his tups, and I didn't want to compound it by letting them make a break for freedom. It was raining quite heavily, to make things worse.

'Can you just hod t'oss for a minute, Mrs Johnstone?' I asked, handing her the lead rope before she had a chance to answer. I admit I was a little preoccupied, my mind racing. Meg was still a bit rocky from the sedative, so it wasn't safe to tie her up. I stood the children and Bill the sheepdog in strategic positions, then chased the tups out of their stable and into the garth, thinking Clive could sort them when he got home. Then I put some fresh straw down in the stable, and I was almost ready to talk to Mrs Johnstone.

'Yer couldn't just pop her in t'stable while I put mi dog away?' I said.

Back in the kitchen it was time to turn to the reason for this visit: the children's progress at school. My Domestic Goddess plans had gone completely awry: the table was awash with hats, gloves and horse equipment.

'Come through to t'living room, Mrs Johnstone, you're looking a la'l bit cold and damp,' I said. 'We might as weel sit by t'fire an' get warmed up.'

Alas, there was no fire. Unattended, it had gone out long since.

The children had done very well, and been very helpful, on a difficult morning. But by now they were getting twined (irritable) and hungry.

'Violet 'ere is lookin' forrad vary much to comin' to t'school,' I said.

'Nah, I'm not,' said Violet, petulantly. I ignored it.

'Violet can count, backard and forrard,' I said.

'Errr, yan, twooo . . . errr.' Then she started to pick at her nose.

I gave up, and listened as Mrs Johnstone gave me a run-down on the academic progress of the older children. They were all performing well, were well behaved, and although some were more enthusiastic about school than others, there was nothing to worry about.

I noticed that Mrs Johnstone was talking nasally, sniffing a bit now and then. She seemed to have a stinker of a cold. This was not good news for me: I have the constitution of an ox with an immune system that can withstand any farm bug, but literally every time I venture into a town or city I come home with a horrid cold which I then share round the family.

'Is'ta badly?' I asked. 'Are yer comin' down wi' summat?'

'No, I've got an allergy,' she said, dabbing her eyes with a handkerchief. 'I'm allergic to animals.'

Oh God! I was mortified. I'd subjected the poor lady to something akin to her worst nightmare. Horses, sheepdogs, tups, and, even as we were talking, Pippen and Chalky were laid out on the rug in front of her.

We talked about the children a little while longer, her eyes reddening and streaming. Happily she soon announced it was time for her to be off. I got the feeling that her trip to Ravenseat had been a memorable one and we'd made a lasting impression, but perhaps not the one I had hoped for.

The vet came back the next day, and we got the tube down Meg a little further, with the same result: some horrible stuff came up, but there was clearly still a blockage. Meg was not distressed, but she still couldn't eat and could take only small sips of water.

The vet asked us to take her down to Sandwath farm, a five-minute drive from the surgery, so that they could attend to her more easily. But she hates travelling, and we knew she would be unhappy off her own patch. Our vet bill would be less, but we decided for her sake to keep her at Ravenseat, where she'd always been happy. We wanted her to stay here for the rest of her life, and not even leave for a few weeks. We had to keep giving her painkillers, but she was so well behaved throughout the whole terrible process.

We didn't seem to be getting anywhere, and she was going downhill fast. After the third day of washing out, I rugged her up in a vain attempt to hide her skinny frame from my own eyes. Her age meant she did not carry a lot of flesh, but now her guts had also sunk right in, her eyes were dull and she looked so tired. Clive and I talked seriously about Meg's future and decided that the next day would be our final attempt, and the vet agreed. He would come to wash her out, but would also bring the humane killer with him.

That morning, I got up even earlier than usual, having not been able to sleep, my mind running over the adventures Meg and I had shared over the years. Over the next couple of hours, a procession of children came to see her. The older ones understood the gravity of the situation. Reuben, as usual, was the pragmatist:

'Does t'digger need its battery chargin'?' he asked. Raven glared at him.

Violet had the stethoscope out.

'Are you checking for her vital signs?' I asked. 'Listen to her guts. If they're still workin' it'll sound like a washing machine.'

'Naw. Can't hear owt,' she said, crouching at Meg's side as she sat amongst the straw and put the stethoscope under the rug.

'Let's go for the heartbeat, then,' I said. Violet moved round to Meg's chest and, after listening intently, shook her head.

'Naw, can't hear 'owt.'

Meg turned her head wearily, and gave a snort.

The vet came, Meg was roused and the tube went down again

for the final time. We watched as bucket after bucket of warm water went down the funnel into her stomach, only to be regurgitated with the familiar smelly food debris. Reuben arrived with yet another bucket which was tipped down the tube and a hush fell as we could hear faint bubbling noises. Seven faces peeped over the stable door, smaller children being lifted by older ones. The vet pushed the tube a bit further in and suddenly a vortex of swirling water disappeared down the tube.

'That's it,' said the vet. 'The blockage has cleared.'

I couldn't believe it; I'd steeled myself for what I'd considered to be the inevitable conclusion of this sad episode.

I can't say that Meg looked relieved. She was still a very poorly horse, and it was going to be a long, drawn-out process nursing her back to health, but I certainly wasn't short of volunteers for nursing duties.

Immediately after scanning we take the yows back to their heafs, giving them their late breakfast. A couple of weeks later we bring them back down into the sheep pens for a bit of a sort-out. Anyone who read my first book, or who knows anything about hill farming, knows about heafs: the sheep, who appear to be wandering aimlessly about the moor when you see them from your car as you drive past, actually know their own territory, the area of the moor that belongs to our farm (and, more importantly, to them). There are no physical barriers to stop them wandering wherever they want, but they have an inbred instinct that keeps them on their own patch. One of our heafs is on common land, where a number of nearby farmers also have heafs and the right to graze sheep. By and large the sheep are bred to know their own heaf and do not stray far off their patch; the yows in turn bring up their lambs to know it, too. When you take on a hill farm, whether you buy it or take out a tenancy, the sheep come with it, and when you leave, they stay behind. We are really just temporary custodians of the flock.

Knowing which sheep are carrying twins means we can feed them extra rations. The yows who are geld are kept separately. They are not neglected and are given a bite of hay every day, but we don't give them any more than that. One or two of them may be in lamb: if they were only in very early pregnancy, it wouldn't have shown on the scan, and they'll lamb late, nearer clipping time when there will be plenty of grass about. The others will go back to the tup next year. As our old friend Jimmy would say: 'A geld yow nivver broke naebody.' They will be bigger and stronger for next year. If they don't breed a lamb for two years running, that's another story: they're off to the auction mart.

Depending on the ferocity of the weather, we may be feeding the flock twice a day, and we watch our hay dwindling, bale after bale going out to the fields, totting up whether we will have enough crop to see us through. It seems as if winter is never-ending, and even though the days are getting longer and we're no longer working in the yard in the dark so much, spring seems a long way off.

We use the hay that is stored in the big barn first, because we need to clear it for lambing in April. It's the easiest barn to fill, a newer, more open building, and we can back the quad bike trailer right in to load it. When that barn is empty we make a start on the smaller, traditional barns: not as easy for loading but we think they are better for storing and preserving hay than the modern ones. They are darker, cooler, and have soil floors.

We once decided that the answer to our winter feed worries was to have a go at growing turnips in one of our fields at Sandwath. Turnips are a good feed, the cows can eat the green stalky tops, and the sheep can eat the chopped-up root. They last well through the winter if you store them in straw, so it seemed a good idea to have a try at growing our own rather than buying them in, and paying a fortune for haulage up to Ravenseat.

Clive's turnip mentor was his friend James, who grows them, and everything else, very successfully on his farm in Crosby Ravensworth, where I lived before I met Clive. According to James, timing is everything, and when it came to planting them he lent us his turnip seed drill.

'When do we do it?' Clive asked.

'June. The 9th is ower soon and the 11th is ower late.' So the 10th it had to be. He told us what to expect from the crop: 'They'll either be t'size o' cannonballs or cat's knackers.'

James had a state-of-the-art turnip-lifting machine to pull them from the ground: Clive had a wife and children. James invited us to a pick-your-own event, lifting turnips from a strip of headland which he needed to clear to get his supersized machine into the field. Only a couple of idiots like us would jump at an offer like that. We managed to fill a whole trailer, and these turnips were definitely cannonballs: one weighed in at a massive 25 lbs.

Another friend of ours found some antiquated, rusty turnip knives in a barn. They probably hadn't been used for years, but with a bit of a rub on the sharpening stone they were soon usable. They had a small billhook on the end which you could use to pull the turnip from the ground, and a sickle-shaped blade to top and tail the turnip, which should have made the lifting easier. But even with the knives, our weekends were spent in the turnip field, bent almost double. And our turnips were, sadly, the cat's knackers variety. The children found it a novelty at first, but that soon wore off. There's an argument, which still goes on, about whether they were turnips or swedes, but the one thing I do know is that when I chopped them up to make chips for the family, they tasted bloody awful. We decided, after that, to revert to buying turnips in.

We have been trying for a few years to establish trees as shelterbelts in a landscape that doesn't afford many places for birds and wildlife to escape the weather. Planting trees has been a labour of love, involving a lot of digging and some manual dexterity planting them on the sides of dangerous ghylls, and fencing them off to protect them. I drew up shopping lists of them, looking through catalogues at all the fancy, exotic species, like monkey puzzles, and the ones with beautiful blossom, like cherries. Then common sense would prevail and we'd order the sensible ones that we knew had a chance of surviving up here: hawthorn, mountain ash and larch.

Our most recent order, in February 2014, involved Clive picking the trees up from a nursery in Northumberland – an errand he really enjoys, as he takes with him his friend Colin, who is familiar with the roads and also with a certain pub along the route.

'I've ordered six hundred saplings, six hundred stakes and six hundred tree tubes,' I said. 'You're gonna need to take the little trailer to get it all in.'

'Nah, we're goin' in t'pickup, we'll get it all in,' Clive said, confidently.

They were away a while, far longer than expected, and then finally I saw them driving very slowly and carefully down our road. The pickup was well loaded – or, to be accurate, overloaded. The tree tubes come in bundles, and having run out of room in the back of the pickup the tubes had been tied on to the roof with baler twine. The trees on the back seat were sticking out of the windows, and others were poking out of the tailgate. The vehicle looked like a camouflaged missile-launcher.

We can't simply decide to plant trees wherever we want: we also have to clear it with an ecologist and an archaeologist. When we wanted to dig a few shallow ponds to encourage wading birds like redshanks, we had to consult them both, and an archaeological survey had to be done on the proposed site. There have been folk living at Ravenseat for nearly a thousand years, but there's little remaining evidence of them, other than old field boundaries and quarry workings. Even so, this is enough to hold up tree-planting and pond-digging. The archae-ologist said, 'There's evidence of peat cutting in field NY8603 2501, where were you wanting to put the ponds?'

'Yeah, yeah, our predecessors were diggin' there too . . . and that's exactly what we're goin' to be doin' . . .'

He ignored me. 'I suggest you build your ponds one hundred metres to the north of the proposed site,' he said, pointing at the map with his pencil.

'Yer want us to build t'ponds on t'steepest part o' t'hill?'

That's what we had to do. It's unlikely that we'll live to see the full reward for our tree-planting labours, but we hope the next generation who farm Ravenseat will appreciate them. We are already a haven for wild birds but anything that we can do to encourage them to nest here and rear their chicks must be a worthwhile undertaking. Only recently we saw a newcomer, a bright green bird which, after consulting the bird book, we now know is a siskin. Clive's passion for birds (the feathered variety) has over the years rubbed off on me. I saw an advert placed by the Royal Society for the Protection of Birds in a farming jour-nal a few years ago, asking for farmers to allow volunteers to do surveys of the birds on their land. All of the birds would be recorded, and the farmer would get a map of the farm with all the species that had been found, and where. The walkers who pass through here ask me all sorts of questions, from the sublime to the ridiculous, but I've always felt I would like to be more knowledgeable about the birds. A map pinned up in the wood-shed for people to refer to would be interesting and informative. So I rang the number, gave them our details, and all was going well until I told the RSPB the size of Ravenseat.

'Ahh, now I'm afraid we can't do an area of that size,' she said. 'And it's not just the size, it's the rough terrain.'

Understandably, most volunteer bird watchers want a stroll through a few pastures and meadows, not an endless hike across bleak moors. It was going to be a major undertaking to map Ravenseat, and for that reason she said that they couldn't do it.

I forgot about it until a few months later, when we were called by Chris, the Conservation Adviser for Northern England for the RSPB. He still didn't have any volunteers, but he said he'd like to do it himself in his spare time. It took him months to complete the map, coming up for a couple of days at a time. He was extremely thorough, and in a way I felt envious of the time he spent quietly sitting and watching what was going on. It is hard sometimes to stop and appreciate the views. If I sit back, I soon notice something untoward, a gap in a wall, a yow limping, or sometimes even an animal's body.

Maintenance of the barns and drystone walls at Ravenseat is an ongoing, never-ending project. Slates blow off during the winter storms, wooden spars and loft floors rot and the weight of snow causes walls to fall and foundations to shift. We can keep up with running repairs but sometimes a barn will need scaffolding and more specialist equipment. That is when we will get the experts in: builders who specialize in the restoration of traditional buildings. The children love these places, dreaming and spinning stories of the people whose names are scratched into the lintels and beams. In summer they plead with me to let them take a pop-up tent and sleeping bags, for a camping expedition inside a barn. But when they hear faint scratches and the scampering of tiny feet, or rain trickles through the fractured slates in the roof, they are back in the farmhouse before the sun rises.

When I first admired the precision stonework and the watershod outer walls of the traditional barns built centuries ago by those who lived and worked at Ravenseat, it occurred to me that they were skilled craftsmen who built things to last. Perfectionists in their art, maybe rather more like Alec than Clive.

It's interesting to discover how much history of the area is actually recorded. It's a question of talking to the right people, and perhaps also a little bit of luck. My friend Rachel was fortunate enough to be given a whole cache of old documents about the area and I spent a fascinating evening sifting through photographs and papers going back as far as the 1820s. I've always loved trying to put faces and histories to the names of people who have left their mark on the place that I call home. Obviously, the further back in history you go the more difficult it becomes.

One afternoon we'd been to pick up a stray yow that had been gathered in by one of our neighbours and left in a fold a way up Sledale. As it was a decent day and the pens were only accessible on foot or by quad bike, I'd loaded the children into the trailer and packed us all a picnic so that we could do a bit of sightseeing and exploring. The children were excited, as we were heading off our heaf and straying into foreign territory ourselves, rather like the yow we were going to retrieve. We decided to investigate the Stone House bridge, an unusual edifice that has always intrigued me. Spanning the river Swale near its source, away from the prying eyes of visitors on the Swaledale road, the bridge is solid, well built, but only connects two fields without even a track leading to it.

I had been told a tale that a father and son from a local farming family had built the bridge to showcase their skills. They competed against each other, building from opposite sides to see who could get to the middle and do the better job. When you look closely at the bridge you can see differences in the walling styles, one side having tighter joints and using smaller stones while the other was walled a fraction more crudely. The children were in their element, running back and forth, from side to side and hopping from stone to stone in the river beneath.

But stories can and do become distorted over time, and it was when I was talking to Rachel that evening that I learned the real story of the bridge. It was not a father and son who'd built it, but two brothers, and they'd erected it in the place where their father had drowned trying to cross the river at night, at a time of year when it was in spate. Indeed their father's body had been pulled out of the river Swale some eleven miles downstream, such was the force of the torrent. Together they built the bridge, a fitting memorial to him, even if it's only used by sheep. One of the brothers, when young, fathered an illegitimate son with the housekeeper, who refused to marry him. The boy left the Dales for London and found himself in great demand as he too was a skilled stonemason and made his fortune, reputedly, building the Holborn Viaduct.

I was also surprised to discover that back in the eighteenth century Ravenseat had its own public house. I knew that our woodshed had previously been a chapel but it seems that as well as the God-fearing inhabitants who refreshed their spirits in the tiny ling (heather)-thatched chapel, there were others who preferred a different type of refreshment. It's hard to believe that this isolated, remote farmstead at the top of the dale was once so busy that a pub thrived here, selling ale to the men from the lead and coal mines, the quarrymen, the packhorse men, as well as the shepherds, labourers and drovers. There was a coal depot at Ravenseat, where the coal from the mines at Tan Hill and the other pits (some of which were working as early as the thirteenth century) would be stored then collected and moved on by packhorse, the same horses that brought supplies of meal and potatoes back to the hamlet on the return journey. In the year 1670, 150 loads of coal came through here: it was a busy place. Nowadays, with the constant stream of coast-to-coast walkers and visitors to the shepherd's hut, I like to appreciate that Ravenseat has gone full circle and is again as bustling with people and life as it was back then.

'Thar ye are,' I said to Clive as I told him these tales of the history of our home. 'Sex, tragedy, all them folks comin' an going: 'twas all 'appening back then just t'same as it is now . . . some things nivver change.'

3

March

Sharp frosts in the early mornings mean that we are still wrapped up in layers of fleeces and waterproofs in March, but there's a vague hint of spring in the air, and later in the day we get occasional shafts of bright sunlight to remind us of the glorious beauty to come.

Old Jimmy was one of our neighbours, a typical Dales farmer of wild and weatherbeaten appearance, his outwardly unkempt attire belying the fact that he was learned and well read and had forgotten more than many people ever knew about the history, sheep and people of this area. Born here in the Upper Dale, he spent much of his life on the moors, and revered a small plant known as mosscrop. He would pick a tiny sprig, put it in the top pocket of his tweed jacket and present his wife Elenor with it, as the first flower of spring. Its emergence in March was for him symbolic: a sign that heralded better weather, and a good crop of lambs. I too look at this tiny unassuming plant with great fondness. For not only does it tell me that spring is just around the corner but it also reminds me of this most enigmatic Dalesman who I was fortunate enough to be able to call my friend.

Our Raven is fourteen now, and a very handy lass to have

around. She's clever, quite academic, very practical and – dare I say it – sensible. From driving the Land Rover or tractor round the fields at hay time, to riding horses, to serving cream teas: she can turn her hand to anything around the farm. She keeps an eye on the reservation book for the shepherd's hut, watching for a vacant weekend when her friends can take it over. The shepherd's hut is a perfect place for a teenage party, they can play music, toast marshmallows on the fire, and talk all night without any interference from me. There's a bonus for us, too: when they stay at Easter they're happy to take a turn at the night-time shift in the lambing shed.

The mother of one of Raven's friends was worried about 'putting on' us.

'Y'know, I don't want to cause you extra work, you've got enough people to feed . . .' she said.

'Don't worry 'bout it. Honestly, if I'm cookin' tea for nine then yan or two more isn't gonna make a happorth o' difference,' I said.

I'm very happy to buy Raven the clothes, shoes and things she wants: I think she's reached an age where these things are more important when she's in the company of her school friends. Back on the farm she doesn't care, and like me she goes for a mix of mismatched garments, mostly charity-shop finds. One morning I was rummaging through the wardrobe looking for something a little tidier to wear when she said, 'When yer die, mother, can I 'ave yer clothes?'

I suppose I could have been upset that she was thinking about my death, but in fact I was quite flattered that my teenage daughter likes my eclectic taste.

Clothes shopping is definitely not one of my pleasures. The little ones don't really care what they wear, as long as it's warm and keeps the rain out. But as soon as it matters to them, I'm happy to get them what they want within reason. I'm not so old I can't remember the agonies of being a teenager . . .

In March, the older sheep that have been sent away for the winter come back to Ravenseat. Sometimes we hire a haulage contractor to bring them all back in one go, or make a series of trips with our trailer if they're not too far away. We need them back in good time so that they acclimatize to life in the hills, and settle down before lambing. We have to check all their ear tags and record their movement. We must stick to the rules and regulations issued by DEFRA and record all details about animal births, deaths and movements, and there is a lot of bureaucracy around it. I can see there have to be controls: nobody would like a repeat of the foot and mouth epidemic of 2001, but at times it seems to be an unnecessary burden. It is easy to unwittingly land yourself in trouble. For example, we once took some heifers to some summer grazing on another farmer's land near Appleby. He doesn't keep cows, and it suited us both for ours to graze his fields; I duly registered on the computer that they had moved off Ravenseat. But then one of them calved, earlier than we expected. She was a fine calf, the heifer an attentive mother. All that I needed to do was register the calf's birth and apply for a passport for her. I gave the mother's ear-tag number, the breed and the date of birth.

Then I had a phone call:

'You've tried to apply for a passport for a calf, but the mother is not on your holding, according to our records.'

My heart sank. I told the truth and explained what had happened: it would have been easier if I'd lied and said we'd brought the mother back here before the calf was born, and claimed I'd simply forgotten to register the movement. I tried to explain that nature doesn't always observe the jurisdiction of the Cattle Tracing System.

'The owner of the land where the calf was born is the keeper of the cattle, and he has to register the birth and apply for a tag and a passport . . .'

'But he doesn't keep cows. He's never had to register one, and it's an enormous amount of work if you've never done it before. Isn't there a way round this?'

'Yes, there is an alternative: put the calf down.'

I was incredulous. It was born in the wrong field, that's all. I had to help the other farmer through all the hoops of registering as the owner of the calf and getting a passport for it, then passing it back to me. I sometimes daydream about the days when cows went by their pet names, and Buttercup was moved from field to field and lived to a ripe old age . . .

Before we changed to Beef Shorthorns we used to breed Belgian Blues, a fashionable breed at the time. One year we bred a wonder bull calf, an amazing creature. He had great length, was nicely marked, stood correctly and had that most desirable feature, one that all breeders aspire to: an unbelievably enormous backside. Everything about him looked good, and we just knew that one day he'd be a show winner.

We decided to use a creep feeder, a hopper with a trough that you can fill with their feed ration of barley and pellet mixture, but designed so that only calves can fit through the graduated bars, not their mothers. They could go back and forth in and out of the feeder without the bigger animals taking any. The calves loved the feed, and the greediest of all was our show-stopper.

'I'll tell yer summat. That calf in't 'alf rattlin' them bars when 'e goes into t'feeder,' Clive said.

'Aye, it's that big old arse,' I said.

With hindsight, we should have altered the bars then, but we didn't. The next day, when we went to refill the feeder, the calf was dead: jammed in the feeder. He had become wedged, panicked and had a heart attack from fright. It felt like we'd bred it to have an enormous behind, then this, in turn, had killed it. We had to prise him out of the feeder and cart him across the bridge for the knacker man to take away.

When you have a flock of sheep or herd of cattle you know that some will die, and you learn not to get too upset by a dead animal. If I've tried my hardest, then I have to accept it. We have our old crusties that die of old age, but there will always be the odd, unexplained death. The downside of a flock that roams the hills is that sometimes a sheep will become ill and die before the shepherd finds it. From time to time one of us comes across a sodden pile of wool in a bog, a plastic ear tag floating on the watery grave. There are times when we find a dead newborn lamb that has suffocated because the birth sac covered his face, and it is maddening to think that if only we'd have been there minutes sooner then we could have saved him. But there is no point dwelling on the 'what ifs'.

A good friend of ours had what he thought was the best tup lamb ever, but unfortunately it died. He still took it to the pub to show everybody what a cracking lamb it had been.

Another friend, a small-time sheep breeder, told us how he had overheard a couple of old boys talking at the bar in the local pub. They were describing the best tup they'd ever clapped eyes on.

'What a heeead 'e 'ad,' said one.

'Aye, couldn't be matched for hair, I'll tell thi,' said the other.

They talked on all night, until our friend tapped one of them on the shoulder and asked where he could see this amazing sheep.

'Nay, t'hell, 'ee's deeead,' said one of them.

'Deeead in t'summer o' 79,' said his mate, chipping in.

Gone, but clearly not forgotten.

We're not allowed to bury dead animals on the farm: we have to have them taken away to the knacker's yard. Considering the line of his work, our knacker man Geoff is surprisingly jovial, going about his business with a smile and taking it all in his stride. He is charming, charismatic and a larger-than-life character. Whenever we see him when he's not on his official duties, he has a pretty girlfriend on his arm. None of his ladies appear to have been put off by his unsavoury job or even the faint but pervasive scent of the knacker van.

His lorry holds a gruesome fascination for the children, who always want to have a peek at the carcasses inside it.

Dealing with dead animals all day hasn't put Geoff off live ones, and he has a menagerie of waifs and strays. He has a passion for peafowl, and when our resident couple, Mr Peacock and Mrs Peahen, finally managed to rear a couple of chicks, we gave them to him. Peahens are not the best mothers: Mrs Peahen would lay a clutch of eggs, sit on them until the first egg hatched and then give up on the rest. She would then trail her lone chick all over the farm until it died of exhaustion, unless I intervened. With her chick dead or removed by me, she spent the rest of the summer harassing the chickens and trying to steal their chicks.

So one year I confiscated her eggs, put them under a broody hen, used a heat lamp after they hatched and soon had a whole chorus of peafowl squawking around the farmyard. It wasn't the charming picture I had envisaged. They became predatory, stalking me when I carried a teatray to the picnic tables, and hanging around the kitchen door hoping for leftover toast from the children's breakfasts. One day they ventured inside to peck at the crumbs when the wind blew through the house, the back door slammed shut and the six of them were trapped inside. Panic ensued as they flew at the closed kitchen window, knocking over a vase and some cups to the floor, losing a few of their feathers in the chaos. What a mess: smashed crockery interspersed with bird droppings.

That was it, they had to go. Luckily, Geoff was happy to rehome them on his smallholding, and that was the end of our peafowl breeding programme. Sadly, Mrs Peahen died a few months ago, and Mr Peacock went into mourning for some time. Once he finally accepted that she had gone, he switched his amorous attentions to the hens. He'd strut his stuff, then try to corner one on the packhorse bridge, blocking its way by fanning his enormous tail feathers. In a moment of madness I asked Geoff to keep a lookout for a new mate for him, so the hens may be off the hook soon.

One day Clive and I were loading the trailer with some fat lambs for the auction. As we ran them into the sheep pens, one of them turned round and charged headlong into the metal gate. Normally the lamb would be shocked, but would turn round and run back with the rest of the flock. But this time it dropped like a stone. We thought it had knocked itself out, but it was dead. It was unthinkable to let it go to waste, so I decided on the spot that we should butcher it.

'Come on, Clive, gi's an' 'and to git 'im into t'barn, will yer?'

Tying the back feet with baler twine, Clive helped me suspend the dead weight of the lamb upside down from a beam. I got out my penknife.

'You'll need a better knife than that,' Clive commented. 'You could ride bare-arsed t'London on that. I'm goin' to t'auction now, so thee's on thi own.'

I was rather hoping that Clive would offer to do the job for me, but he didn't, and I was determined that I would see the task through.

You don't realize what a skilled job butchery is until you try it.

Skinning it wasn't a problem: we're good at that, as we often have to skin a dead lamb or calf to persuade the mother to adopt another one. I know how to gut rabbits, so I could do that bit. But then things got complicated. A leg is a leg, and neck end is obvious. But working out where to cut for chops was tricky. I got there eventually, but we had to be careful when we ate it, because lurking amongst the meat were quite a few splinters of bone.

We do, occasionally, take animals directly to the abattoir. From time to time we'll keep a lamb for ourselves, especially if the prices at the auction are not too good. Recently we took two rogue shearlings to the abattoir. They were very fat because they were forever ratchin', which means they would jump any wall or gate, and were always trespassing in the best fields, eating the best grass. Finally they took one jump too many, out of my pickup and into the lairage, the holding area at the abattoir.

We use a very small abattoir where we know Joe, who runs it, very well. We rear a couple of pigs every year, and he does them for us, too. I like the fact that we get everything back, neatly butchered for us. I hate waste, and I use the belly flaps, kidneys, liver to make pâté. Joe will make black puddings, sausages, whatever you like, as fancy as you like. When you eat it, you know its provenance one hundred per cent.

Although we are a sheep farm with a small herd of our own cows, we also rear up cattle that we haven't bred ourselves – buying them in as calves and then selling them on as 'store cattle' to farmers, who will fatten them up for market or breed from them. How long we keep them is dependent on the trade at the market, there's no hard and fast rule. If year-old beasts are selling well, we'll take them to the auction; if not, we keep them longer. The markets are volatile, prices fluctuate, and we can never predict how it will go. But one thing is certain: people will always need to eat.

You have to be a bit careful working with young cattle, especially the bulls. Once they get to about nine months old, the sap starts rising and they can get frisky. Once they are speaned, weaned off milk, they are fed a barley and pellet mixture. In the summer they're out in the pastures, but during the winter months they're fed in troughs in the barn. I feed them twice a day, jumping over the cow barrier with a sack of feed. They are creatures of habit and always know when it's time for their meal. If I'm late, they kick up a ruckus and bawl until the food arrives.

One March morning, I was in the barn feeding the young bulls when one of them started to get a bit lively. He'd gallop up and down the barn taking big jumps, twisting and spinning round, then come to an abrupt halt next to me and try to rub his head on me. I was very wary of bending over the troughs.

'I'm not 'appy goin' in wi' t'stirks,' I said. 'Yan of 'em is actin' a bit stupid.'

''E's just a bit full o' 'imself,' said Clive. 'Just tek a stick wi' thi.'

So I carried on feeding them, day after day, with one eye on the antics of the cavorting young bull. One morning I noticed a wet, muddy patch around the water trough, which had clearly sprung a leak.

'There's an 'ole in t'watter trough,' I told Clive. 'Needs lookin' at.'

'I'll sort it,' Clive said.

I went into the adjoining barn to fill a barrow with straw, while he went to the tool shed to find all he needed: a small nut and bolt, a couple of home-made rubber washers crafted from an old inner tube, and an adjustable spanner. Plus Raven, his unwilling plumber's mate. They went to the far end of the barn, where the trough was. The metal at the bottom had rusted and become paper-thin, and there was a small pinprick of a hole which, if not sorted, would only get bigger. Clive rolled his sleeves up and was peering into the icy water whilst Raven, distracted, was gazing at the peacock perched in the rafters above. The stirks, having finished their food, were watching with interest. The lively one was getting closer than the others, intermittently licking the back of Clive's jacket, for which he got the occasional tap on the nose.

'Bloody 'ell! Ah've dropped mi flamin' washer into t'bottom of t'trough. Mi 'and's so cold I think it's gonna drop off,' said Clive, as he fished around in the icy water. 'Ga' an' git me another bit of t'inner tube, Rav. 'Urry up and stop gawpin'.'

Off she went, leaving Clive scrabbling about in the water.

I heard the noise from the other side of the yard. It was a mixture of hoots of wild laughter from Raven, some serious swearing from Clive and a strangulated mooing noise. I ran round the corner to see Raven doubled up, laughing so much that tears ran down her face. Clive was also doubled up, his hair and face wet. It seemed that Clive had taken a dunking in the water trough when the bull mounted him.

'By 'ell, I din't see that one comin',' he gasped.

'I told yer so,' I said, trying not to sound smug.

The next day the vet was called, and the bull became a bullock.

Sheepdogs are an important part of our lives. Because our ground is so rough and steep in places, we aren't heavily mechanized: the quad bike is the most useful bit of kit we have. But still much of our land is inaccessible except on foot, so we are a 'dog and stick' kind of farm, and we've had some top-notch sheep-dogs over the years. Working dogs tend not to live as long as family pets, but they do have a more natural and down-to-earth type of life, out working on the moors and in the fields every day, come rain or shine, then sleeping in a straw-filled box in a kennel in the barn. They are not pampered but loved, cared for and respected, because without them our job would be impossible. A sheepdog will share the good times and the bad with unwavering loyalty. Some are better than others, and some that we acquire just don't make the grade.

One day I was chatting to Alec, who is an expert trainer of sheepdogs. He was about to set off to the sheepdog sales at Skipton. It was a big event, where he would meet up with other members of the dog-running fraternity. There would be between fifty and a hundred sheepdogs there: some fully trained and broken, some partly broken, and pups up for sale. Broken dogs would be run on a small flock of sheep immediately before their sale, so potential buyers could see them in action. Then the auctioneer, standing on a makeshift rostrum, would set off the bidding. Before Alec left I made a casual remark.

'If thoo ever sees a dog that looks like mi job, would ta buy it for mi?'

Unwittingly I'd just set him another challenge. He rang up a few hours later. 'I've bought thee a dog . . .' he shouted, still getting used to the idea of a mobile phone.

I hadn't expected him to find my perfect dog so quickly, and started to worry, as I knew that the highest price ever paid for a sheepdog had been at the Skipton sale. That dog had made five thousand guineas. Alec didn't reassure me when he boomed down the phone, 'Tha's gonna 'ave to sell a few cream teas to pay for this 'un.'

'Alec, yer 'aven't brokken t'record sale price, 'ave yer?'

'I just saw t'perfect dog for thi so I kept flappin' mi 'and. I tell't 'em to book it down to thee.' I winced.

'Just tell me about it when yer get yam, Alec,' I said, fearing the worst.

He arrived with Kate, who was eighteen months old and fully trained. A little green, perhaps, but she was good, very keen. What she needed was work, lots of it, to bring her on. A year at Ravenseat has brought her on enormously, and Clive's very impressed with her, but she is my dog. People have commented that they even see similarities in our personalities, Kate being a full-on action dog, sometimes even referred to as a bit of an air-head . . . But the price? Three thousand guineas. A lot of cream teas, for sure. But what is a shepherdess without a dog? If she turns out as good as we think, she'll be more than worth it.

In sheepdog trials, it is an absolute no-no for a dog to have contact with sheep, but in real life, as a working dog on a farm, they sometimes do. My first sheepdog, Deefa, was not by any stretch of the imagination the greatest gatherer of sheep; her speciality was catching them. When I'd failed in every other attempt to get hold of the yow or lamb I needed, I'd yell 'Catch it!' at Deefa, increasing the volume of the command until she'd separated the sheep from the flock, knocked it off balance and pinned it to the ground, never harming it, until I caught up. I would lavish her with praise, and she'd do laps of honour round the field in sheer delight. Bill, Clive's dog, won't harm sheep either. He'll snap and lunge at those that refuse to move, some-times grabbing at the wool; but he never bites or punctures the skin.

Bill is very hard, a man's dog. In many ways he's uncouth: he'll wait until there are visitors then he'll scratch himself, fart, and for some reason he loves to poop on the highest thing around – including, on one occasion, on top of one of the picnic benches. Luckily there were no visitors there eating cream teas at the time. He likes playing dog skittles: he runs around other dogs in ever decreasing circles, then comes at them at full tilt, knocking them off their paws. Kate is wise to this, and lies down whenever she sees him approaching. Pippen and Chalky also roll over when he comes at them, but our third sheepdog Fan never learns, and gets taken out every time.

Bill's powerful, not just in his build but in his eye, his personality, his determination. And he's clever. There are two sides to his character. With Clive he's really butch, always runs alongside the quad, responds instantly to Clive's commands. With me he is more relaxed, doing things he would never do with Clive, like riding in the footwell of the quad, putting his head on my knee for a stroke and giving me an idiotic sort of dog smile. He even rolls on his back for a belly scratch. The only time he rolls when Clive is around is if he's found some fox poo, or the rotting corpse of a small dead animal. He doesn't prefer me: he's definitely Clive's dog. But he knows I'm different, and he's not stupid: he knows if he is gentle round me and the children, it will get him further in life. With us he is a big, slightly whiffy ball of gentleness. When little Annas was eleven months old she was with the other children playing down by the river. We'd all been holding her hands and trying to get her to stand up, but it was only when she put her arms around Bill's neck that she pulled herself up, and took her first steps alongside him.

When it comes to handling sheep, Bill knows what you want him to do even before you have time to tell him. It can be a problem when he makes assumptions. He knows the lie of the land, which way the sheep run and where the gates are, so if you want him to do something a little different from normal you end up battling to persuade him you really mean what you are saying. He will set off on an outrun around the sheep, but will keep glancing over his shoulder, looking back with questioning eyes as if to say, 'are you sure you want to do it like this? This is not how we usually do it.'

Our working bearded collie, Fan, is mine. She'll never be a top work dog – she's far too timid to make the grade – but she has a good, kind temperament. She comes to feed the sheep with me in winter, but she does not have the inclination or power to be a good moor dog. She's quite lethargic and laid-back, and there are times when you need speed. She is the opposite of Kate, who is rather more gung-ho and has to be encouraged to 'take time'. Dogs, like people, are individuals, and for a shepherd the relationship does not always run smoothly, but over the years we've had some really special dogs.

Clive's very first dog was one that still holds a really special place in his heart, for sad reasons. She was called Nellie, and she was given to him when he was seventeen by Ebby (Edward Metcalfe), the farmer and dog trainer who was Clive's mentor.

'Look after thi dog, yan day it'll mek a good 'un,' he told Clive.

Sure enough, she was a lovely dog: loyal, devoted and exceptionally clever. Clive was living at home, and every morning when he got up the dog would be lying on top of his old white Cortina to make sure he didn't go off to work without her. He even took her on dates with girls.

'She nivver cramped mi style, quite the opposite, they allus took to 'er. It worked a treat . . .' he'd tell me, smiling.

Clive dreamed of one day being a farmer himself, so in order to raise the capital he needed he worked as a builder for a while – and Nellie went with him, pottering around the building site all day and sharing the sandwiches from his bait box. When he was working on a site only a couple of miles from home, Clive used to run there, and Nellie ran with him.

He was only nineteen, working hard and playing hard. Friday nights were about parties, the pub and girls. At that age, you don't always think things through. When work finished one Friday afternoon, Nell wasn't there. Clive looked everywhere: it was odd, because she never wandered. He set off home, hoping she might be there. She wasn't, so he jumped in the car and set off back to the building site. He found her body on the road: she'd been knocked down crossing a road, on her way back home. She knew where she was going: she was trying to get back to him.

He gathered her up, putting her on the front seat, where she had always travelled, and took her home. He was heartbroken. He couldn't even bear to dig her grave: his brother Malcolm buried her. Clive says, 'I was very lucky to have her as my first dog. Some fellas spend a lifetime and nivver 'ave a dog like her. If yer first dog isn't a good 'un, you could lose heart and mebbe spend the rest of your life doing summat else, like building. But Nell was so good, she was a big influence on whether I carried on with sheep, she got me going. I learned a very hard lesson that day.'

The stories of my dogs, Deefa and Red, and of Clive's dog Roy, are told in my first book. When I moved in, Clive was worried whether I would be acceptable in Roy's eyes. Roy, rather like Bill, had two sides to his personality, and could occasionally take a great dislike to somebody. Luckily I was deemed acceptable in Roy's eyes: who knows what would have happened otherwise?

Sheepdogs come in all shapes and sizes. We prefer a smoother-coated dog, because when they've been working on the hills their coats dry much quicker, which we think means they are less prone to rheumatism as they get older. A dog with a huge, woolly coat can still be damp the next morning. The tail should be carried low and straight, as this gives the dog good balance. If you see a sheepdog with its tail stuck up in the air, chances are it will be a muppet. Clive says he likes his dogs like his women, 'On a longish leg, and fast.'

I say I like my dogs like my men, 'Devoted, athletic and dependable.'

Dogs command more respect from sheep if they are darker-coloured, as sheep sometimes chase a light-coloured dog, probably mistaking it for another sheep or a lamb. Deefa was lighter, and occasionally she had this problem, but overall she worked very well, and Clive came to respect her.

After my beloved red dog, Roger Red, died, I asked Alec to look out for another red and white collie. They're quite a rarity, and Roger had been a character who served me well, so I thought another red one would fill the gap he left. I had been to look at various pups that had been advertised as red and white collies, only to discover that they were more of a brindled or roaned colour.

Alec eventually found me a lovely pup I named Red, in honour of his predecessor. He grew to be a very handsome dog, big and beautifully marked, with a sleek, shiny coat. He was unswervingly loyal, would go everywhere with me, do anything for me. Well, almost anything. There was one thing he wouldn't do: he wouldn't chase sheep. He wasn't frightened of them, but there was no way he would run after them. I tried unsuccessfully to get him interested, chasing amongst the sheep myself with him and even bringing the other sheepdogs down to the dog training field to provide encouragement. Surely he'd want to join in? No, he just stood beside me, surveying the scene, wearing an aloof expression.

A nutcase of a dog that wants to chase sheep can be trained, its instinct to chase honed and its desire to worry the sheep curbed. But a dog that shows no interest in sheep is impossible to train.

I gave Red every chance to be the dog that I needed, but it was not to be. In the end, I decided to ring a sheepdog charity and ask them if they would be able to rehome him, explaining that he was striking to look at and had no behavioural problems. I was sorry to part with him, but a working farm needs working dogs and Red was destined to be a pet. I handed him over with a clear conscience, knowing it was the best thing for both him and me.

Four or five years later, on a sunny afternoon, I was busy going back and forth between the farmhouse and the picnic benches serving cream teas to walkers. I was chatting away, and then I noticed one couple had a beautiful red and white collie. I stopped and admired him.

'Smart lookin' dog,' I said, nodding at the obedient pooch lying at their feet.

I reminisced about my red and white sheepdog, how I'd once been given a good one called Roger and how I'd always wanted another . . . As I babbled on, I looked at their dozing dog. I was about to talk of Red, the sheepdog that never was, when I stopped short.

I know you, I thought as I studied him. He raised his brow, one eye looking up at me without moving his head, which rested on his outstretched front paw. I thought I detected a flicker of recognition.

Back in the farmhouse, I talked to Clive as I put up the order: 'Yer know what, I's thinkin' yon dog down there is t'red dog that wouldn't work.'

'Nivver! Is ta sure?' He sounded incredulous.

We both went down with the cream tea order. Clive studied the dog on the quiet.

''Tis a nice dog,' I said to his owners.

'Yes, he's a good 'un. Fit as a flea, we do a lot of dog agility with him.'

They told me they lived at Penrith, not a million miles from Ravenseat. I quizzed them further.

'Is 'e from a workin' stock?'

'Aye, he's come off a farm,' said the man as he poured his tea. 'He was a rescue case, been abused, come off a terrible place, always tied up, never let out . . .'

I didn't say anything; there didn't seem any point. The end result was good: Red had found a loving home.

'That was mi red dog, weren't it?' I said to Clive as we walked back up to the house.

'Wi'out a shadow o' doubt,' he said. 'I nivver forget a face.'

We also had a dog called Nip, a big black dog with pricked lugs who'd round up sheep, no problem, but the minute your back was turned he'd go back to the farm. Working near to the farm buildings, in the confined space of one of the fields, this wasn't a big problem, as you could see what he was doing and block his exit. But at the moor, in the company of other shepherds with well-trained dogs, it was a different story. You could end up looking very amateurish when your dog hightailed it home without you. We gave Nip to a friend who farmed on a smaller scale, and they worked exceptionally well together. It's all about matchmaking: the right dog for the right owner.

Dogs do, on occasion, go missing; it's an occupational hazard. Our worst nightmare is when we hear that a sheepdog has gone missing, because we know that it will instinctively want to chase sheep. A dog loose and out of control at the moor can create a heck of a mess, chasing sheep off their heafs and causing mayhem.

Sheep worrying, although increasing across the country, has never really troubled us at Ravenseat, as we are so remote, and the sheep roam such large areas that there is little interaction between them and the walkers' dogs. But we did have an unsettling episode of sheep worrying when we kept some sheep in a field at Kirkby Stephen, the one we use as a car park for the annual Cowper Day horse fair. There's a narrow lane that runs alongside the field called Bloody Bones Lane (which some local historians say got its name from a massacre of Jacobeans invading from Scotland; more prosaically, others say the name derives from the days when butchers used the route to dispose of carcasses out of town in medieval times).

The only people who used the dark tree-lined lane were, generally speaking, ramblers, dog walkers and us, when we went to tend our sheep. Robert and Alec kept an eye on them, and gave them a bite of feed each day.

Early one morning we had a phone call from Robert to say that when he went to feed them he found one of our sheep had been badly mauled and killed. We went down to see, and I really wished that I hadn't – it was a horrendous sight. The dog had clearly cornered the sheep and gone for her face. There were deep bite marks to her throat, and I grieved for the pain and terror she must have felt. The culprit was clearly a dog, for neither a fox nor badger would be capable of such savagery to an animal of this size. I took a photograph, which was made into a poster and put in the window of the local vet's, asking dog owners to watch out for irresponsible behaviour, or lone dogs without leads roaming the area.

We soon heard that it wasn't an isolated incident, as other sheep belonging to farmers in the area had also been attacked and worried. Was the dog loose on its own, or was its owner with it? Was it being done deliberately? We walked through the field looking for paw prints, bits of fur, but we found no clues. We hoped we'd just been unlucky, and that it was a one-off.

The following week, on the same night of the week, another of our sheep in the same little flock was killed in the same horrible way. This time I put a picture up on Twitter, with a warning that it was a gruesome sight. The response was lots of sympathy, but no useful information. Clive and I talked about it all the time. What kind of dog was it? A terrier? A lurcher? Some folk suggested that there might be a big cat, like a black panther, roaming the neighbourhood: one had apparently been sighted up the back lane near the auction mart. Of course, such stories were met with a great deal of derision by most. It was only when the panther was sighted again, this time reclining on a tree branch by the woods, that people took this possibility seriously. Supporting evidence in the form of some very large scratches in a nearby tree trunk, and some sizeable paw prints, added weight to the escaped big cat theory.

We lost another sheep in the same way, but this time on a different night of the week, so it was impossible to work out a pattern. Our tiny flock was fast dwindling. We reported it to the police every time, but there's not a great deal they can do without information about who, or indeed what, is doing it.

So that's when Clive and Alec decided to take matters into their own hands. It's a blokes' thing. They borrowed a wildlife camera, which had a sensor triggered by movement, and set it up in the field, but unfortunately it didn't really work as planned, because there was always something moving: the sheep. So when they went back to check the next day, all that had been recorded was the green reflection of the sheep's eyes and the occasional owl flying past.

It was back to the drawing board, and another plan was

hatched. Clive and Alec were going to stake out the field all night.

'We're gonna tek it in turns an' just sit an' wait,' said Clive.

'Yer can count me in,' I said. I was having sleepless nights worrying about it, so I might as well spend them doing something constructive.

'Then what?' said Clive. 'What yer gonna do if summat 'appens?'

'I'll tek mi camera, get a picture, an' ring the police. Anyway, what's thoo gonna do?'

'Well, you'll mebbe shoot wi' yer camera, but I'm gonna shoot any dog I see in mi field wi' mi shotgun,' he said defiantly.

'Great! So we're gonna end up with a firearms incident, too . . .' I said.

'That's nowt,' said Alec, 'I'll shoot bloody dog and anybody tha' I see in t'field.'

Even better. But I was sure it was bravado, and doubtful they'd be as brave if faced with a snarling dog and an equally snarling owner.

The plan was simple. Armed with a high-powered lamping torch borrowed from the gamekeeper who lives in the cottage behind ours, we would park the Land Rover in the field gateway at the top of the lane, partially hidden by trees and bushes, and just sit tight from nightfall until dawn. Raven, Reuben and Miles were desperate to take part in the undercover surveillance, so we chose a Friday, when they didn't have school the next day, for our stint. We packed the essentials – chocolate and a flask of strong coffee – drove down there and settled in, with a subdued sense of excitement as we peered out into the darkness. The hushed talk of wolves, spectral hounds, big cats and marauding sheep rustlers soon wore thin, and after hours of absolutely nothing happening, the children fell asleep. I was so bored I had to force myself to stay awake, and we were all happy not to volunteer again. It was much the same for Clive and Alec, who covered the rest of the week between them. We were all tired, and spirits were low. Then a chance encounter ended our night watches.

Apparently Clive, after sitting quietly for a few hours, decided to get a breath of fresh air and stretch his legs. His eyes were adjusted to the darkness, and he could see quite well as he walked along the lane. He paused, momentarily, thinking that he could hear something. Then he decided to have a piddle against the hedge before going back to the Land Rover.

Just as he got back into the driver's seat, he saw the blue flashing lights of a police car edging up the lane. He was heartened: they were clearly patrolling the area and taking the attacks on sheep seriously. An officer walked up to the Land Rover and signalled for Clive to wind his window down.

'Evenin', officer,' said Clive.

'We've had a report that there's been a sighting of someone acting suspiciously in the bushes,' said the constable. 'Someone answering your description . . .'

After that we gave up our stakeouts, and there was no more sheep worrying. We are no wiser as to who or even what was responsible, or why it stopped.

One dank, depressing afternoon in March I left Ravenseat to pick up some shopping and visit a neighbour in hospital. On the way home I made a small detour and called in to see some friends for the evening. It wasn't a riotous night out: just lots of tea and chat about sheep. I didn't think much would be happening at Ravenseat. But I was wrong.

After tea Miles had sauntered out to look at his Texdale shearlings, which we'd put in the barn as they were first-time mothers-to-be and were, by now, getting close to their due date. He came running back to the house with the news that one of them, known as Fatty, was making funny noises – she'd been scanned for twins.

Clive went outside and confirmed Miles's diagnosis, but he soon realized that there was trouble when three mismatched hooves appeared.

'Aw, what a tangle-up there was,' he told me later.

It's not unusual for us to have trouble at the very start of lambing time. Miles didn't grasp the seriousness of the situation and knelt by Fatty's head, stroking her while Clive tried to push back the unborn lambs.

'It's naw good, Miles. We need to get 'er to t'vet,' he said, standing up, looking down on the outstretched, straining sheep and shaking his head.

Fortunately, I'd taken the pickup and left Clive with the Land Rover, into which he loaded seven children and a groaning sheep. As he drove as fast as he could to Kirkby Stephen, he prepared the children. He didn't mince his words.

'Kids, I don't reet know what t'plan is, but I doubt we'll be getting much joy. I ain't done much at 'er, but I think that t'lambs will be dead.'

We don't shield the children from the harsh realities of our lives, and Clive didn't want to raise their hopes.

Lesley the vet was more upbeat. 'We'll sort this, nae bother,' she said.

Clive backed the Land Rover up to the operating theatre for large animals, and Lesley set to work.

'Hod this,' she said to Edith, handing her the Vet Lube (a lubricating gel).

'Hod that,' she said to Miles, giving him the end of a lambing rope secured to one of the lamb's legs.

Clive held the poor sheep while the other children watched. After a bit of groaning and sighing the yow delivered a decent-sized gimmer lamb, alive and kicking. Raven took charge of the newborn, rubbing her vigorously and squeezing her nose to clear the birth fluids. Moments later came a tup lamb, alive, considerably bigger than its twin.

Lesley looked pleased with herself, and rightly so.

'Thoo's done a proper job,' said Clive.

'Well, she's not had an easy time of it. We'll give her Pen Strep (antibiotic) to make sure she doesn't take badly.'

Loaded back into the Land Rover, it was a more leisurely drive back, with tired but happy children and a relieved sheep, nuzzling her lambs. While Clive and Miles built a pen for the sheep, Raven and Reuben supervised showers and a good scrubbing for all the children.

Clive and Miles had a discussion about the two lambs: after the traumatic birth, it was probably too much to expect her to rear two lambs. The plan was to take one from her and bottle-rear it until a foster mother could be found.

'I'll look after it,' Miles said. He liked the idea of bottle-feeding one of his own lambs.

'Aye, I'm sure you will, Miley lad,' said Clive. 'But we'd better bring it into t'warm, and see if Mam's got any colostrum in t'freezer; it'll be better than pinchin' what la'l bit yer yow's got.'

So the lamb came into the farmhouse. First stop was the bottom of the shower, as she was still wet with blood, testament to her difficult birth. Sidney was in the shower, covered in soap suds, his eyes screwed tight shut.

'Ooh, yammy in shower,' he said, when he opened them to see his new companion. 'Yammy clean.'

She certainly was. A blast with the hairdryer, and she was then snuggled down in front of the fire. There were plenty of offers to take her to bed, but Clive was firm. It was time to give me an update.

'How's ta doin'?' he enquired on the phone.

'I'm fine, settin' off yam shortly. Wha's thoo up to?'

'Well, the children have gone to bed but I'm not alone . . .' I could hear a faint bleating.

''Ave you got a sheep in there? What's been goin' on? The minute my back's turned . . .'

He told me about the night's events, and I was soon back home, perched on the fender in front of the fire, bottle-feeding freshly thawed colostrum to the new arrival, and thinking to myself that there was no place in the world I would rather be.

Traditionally, March is the time of year when folks talk about spring cleaning – and talk is usually as far as it gets here. There may be a general tidy-up and a clear-out of unwanted furniture and accumulated rubbish, a half-hearted attempt at making some semblance of order in our chaotic but happy house. There isn't usually much that is salvageable, after the rigours of life at Ravenseat. But one year I decided it was time to say goodbye to a corner cupboard that I didn't like, but which was a bit too good to become firewood.

I took it to a monthly furniture sale that was held in the village hall at Hawes. I have a bit of a history with furniture auctions, usually defeating the object of having a clear-out by buying on impulse more stuff than I get rid of. Sure enough, after unloading the cupboard I rummaged around and spotted a battered music case under a table, among a selection of miscellaneous bits and pieces. I opened the case, and there was a trumpet. I didn't know what sort of trumpet it was; it definitely wasn't a bugle, and it didn't resemble a cornet either. What I could see was that it had been well used, was tarnished, and the valves were sticking. I'd just recently seen the film *Brassed Off*, about a Yorkshire colliery brass band, and momentarily I pictured myself being the next Louis Armstrong. So I noted its catalogue number and left a bid of £30 on it.

I didn't check with the auction for a few weeks, expecting to just get a cheque for the cupboard in the post. It was time to give them a call.

'Did yer sell mi cupboard?' I asked, a little annoyed that I'd had to chase it up. ''Ow much did it mek?'

'I'll have to go through the sale day papers,' said the auctioneer. 'We don't 'ave a computer.'

After a long pause he said, 'Yer cupboard made £220.'

This was good news, far more than I'd expected.

'Excellent, can you put t'cheque in t'post?' I asked.

'Nay, we pay thee when thoo's paid us and you ain't paid for t'flugel'orn yet.'

So it wasn't a trumpet, it was a flugelhorn. And my £30 had bought it.

When I got the flugelhorn home, I had a tentative go at playing a few notes. Way back when I was at school, I'd had lessons on the French horn. I'd been resentful about having to play it, because I desperately wanted to play the cello. I had visions of being the next Jacqueline du Pré, a slightly troubled soul, sawing away with great passion and talent. But Mother said it wasn't ladylike straddling an oversized guitar, with the endpin making holes in the carpet. So the French horn it was, and weekly lessons from Mr Brompton, a colonial type with silvery grey hair and an oversized, wiry moustache. He struggled unsuccessfully to teach me embouchure, taking the mouthpiece from the French horn and buzzing on it. I was transfixed by his moustache: it would twitch as he blew, and minute globules of spittle would fly through the air. Then I'd be told to do the same.

To make matters worse, the French horn was bulky and too heavy for me to carry to school, and a tartan shopping trolley – of the type old ladies sometimes use – was customized for transportation purposes. A letter from the council offering me a place in the brass section of the youth orchestra at Huddersfield Town Hall was the final straw. I was not going to be seen lugging a tartan shopping trolley through the town centre. My refusal was compounded by the rollicking I got for cleaning the French horn with Brasso, and when my younger sister Katie was allowed to learn the cello, I decided enough was enough. I dumped the French horn on Mr Brompton's doorstep, shopping trolley and all.

Reuben soon had the flugelhorn dismantled and the valves oiled and working properly, and he decided that he would like to learn to play it. Brass lessons were available at school, and with no neighbours to annoy with his practice, it seemed an excellent idea – and an even better one when the Muker Silver Band offered to take him on as a new member. Clive is very happy to take him to band practice at the Reading Room in Muker every Monday night, while he spends a couple of hours in the pub talking tups.

There's a long tradition that the band plays hymns outside the Farmer's Arms pub after Muker Show, and Reuben is hoping that soon he'll qualify to wear one of the bright blue nylon band uniform jackets and play alongside the others.

It wasn't long before Miles took an interest in the talk of the mysterious Reading Room and, particularly, the halves of shandy and packets of peanuts shared in the pub after practice. Norman, the band leader, appeared at the farm one day with a cornet for Miles, so he, too, has joined the Muker Silver Band; between them they've probably brought the average age down by about fifty years.

Violet is also showing signs of being musical. She tinkles around on our old out-of-tune piano, and Edith has taken up the ukulele. Whether we will become the modern-day Von Trapps and whether any harmonious tunes will ever emanate from Ravenseat, I just don't know; only time will tell.

From a very early age, Sidney has had his own little preoccupations, and so far, at least, they haven't included music. At one time, he was obsessed with keys: there was nothing he liked better than finding a set. He'd home in on them, swipe them, grow bored and eventually leave them somewhere random. I once spent an entire afternoon looking for the house door key, then found it in his nappy. After that, we instigated a strip-search the moment any key went missing.

We tried to remember to put keys out of his reach, but this led to other problems when one of us would have a set of vehicle keys in our pocket when we set off from the farm, forgetting the other would need them while we were away . . .

On one occasion, though, Sidney's magpie instincts saved the day. It was a bright but chilly afternoon, and despite the temperature I'd had a group of walkers call in for teas. As usual Chalky had decided to linger around the picnic benches waiting for the right moment to leap on to the table and make off with a scone. I could read her mind, and being thoroughly fed up with having to provide replacement scones or give refunds, I decided to thwart her by shutting her in the nearest confined space. She was temporarily incarcerated in the front of the pickup until the walkers finished their teas, packed up and headed off on their merry way. Only then did I go to let her out.

'C'mon, Chalks,' I shouted, as I walked over to the pickup. As I approached, her head popped up at the driver's window, her small paws resting on the window ledge. Then she neatly put one of her paws on top of the door lock, triggering the whole central locking system.

'Good move, Chalks,' I muttered, heading to the house for the keys. There were no keys hanging on the nail in the beam. Probably in Clive's pocket, I decided, so I set off to find him. No, he didn't have them.

'Yer don't think that t'keys could 'ave bin in t'ignition?' he asked.

My heart sank. I remembered I had been the last one to drive the pickup, and I routinely left the keys in. We peered through the window: there weren't any keys in the ignition. Chalky wagged her tail furiously, probably wondering what kind of game we were playing.

'Spare key?' said Clive.

'In yer dreams,' I said.

'We'll 'ave to brek t'window,' said an excited Reuben.

Reluctantly, Clive and I were forced to agree. Just as Reuben was heading off to the workshop to get the mini crowbar, a grubby-looking Sidney emerged from the coal house. I wasn't in the mood for any more problems, and spoke sharply to him.

'Yer mucky lal' bugger, wha's ta doin' playin' in there? I need to get thi into t'bath . . .'

It was then I noticed he was clutching something in his small, filthy fist.

'What's ta got in thi 'and, Sidney?' I said.

Shuffling a little, and looking guilty, Sidney opened his hand. The pickup key!

I swept his sooty body up into my arms, removing the key from his hand. Sidney had come up trumps, and the only disappointed person was Reuben, who had been looking forward to smashing the window.

This was one of the rare occasions when Sidney's key obsession paid off. Mostly, it just caused problems. One bitter cold morning, I woke up to a reasonable (I thought) day ahead. It was a simple assignment. A photographer was coming and we were going up on the moor to take a picture of a hill shepherdess (me), a sheepdog (Bill) and a flock of sheep. The children were all wrapped up. We loaded the photographer and all his gear into the bike trailer along with the little ones and drove up to the most godforsaken, windy, exposed place on the moor, where the views are magnificent but the wind chill factor is extreme. On the way up Bill rolled in something putrid, and when we got there the photographer decided he wanted me dressed in something more summery. Stripping off my hat, jumper and layers of waterproofs left me freezing. And all the while Clive was complaining because, once again, his least favourite sheep kept getting herself to the front of the flock.

'Get tha' ugly bugger outta t'picture!' he shouted.

'Who exactly are yer talkin' to?' I yelled back, to make myself heard above the wind that was whipping around as I shivered in shirtsleeves.

'I hope thoo's gotten a goosepimple filter on thi camera, 'cos it's proper starvation with nae kitle on,' I shouted to the photographer, who was well wrapped up and likely couldn't hear what I was saying anyway, owing to the woolly hat pulled firmly down over his lugs. I was beginning to sound like a real diva, but with a stinking dog downwind of me, Clive being awkward about the sheep, and the children getting very bored and cold, all I wanted was for the photoshoot to be over and to get back home.

Clive was clearly feeling the same way, because after allowing the photographer what he felt was plenty of time to get the shots he wanted, he announced, 'Reet, that's it. No more fannying around, we're going yam.' I think that's the Yorkshire version of 'It's a wrap.'

A family portrait. From left to right: Miles, Sidney, Clive, myself with Clemmy, Edith, Violet and Reuben. Raven is seated with Annas on her lap.

A shot of Ravenseat in the snow. The beauty of the landscape never fails to take my breath away.

OPPOSITE PAGE TOP The winter terrain can be tough and unforgiving for humans, sheep and quad bikes alike. MIDDLE Luckily the sheep cannot get enough of their winter rations! BOTTOM Edith finds another use for our tea trays in the snow.

LEFT February brings Adrian, the scanning man. The children enjoy watching him work, even if their role in the proceedings has been reduced! From left to right: Edith, Violet, Reuben and Miles. ABOVE Annas and her new friend. BELOW We like most of our sheep to lamb outdoors – it's natural and healthier.

ABOVE Alec, as well as being a close friend of the family, is an expert sheepdog trainer. RIGHT Annas plays with Kate while Bill keeps watch. Our dogs might be workers, but they still find time to play. BELOW Overhanging cliffs at the entrance to the Boggle Hole, which is only accessible when the beck is just a trickle.

ABOVE Miles and Sidney love their chickens. Miles is particularly devoted to his hen duties, come rain or shine.

LEFT Edith picking wild flowers. The children love being outdoors and exploring.

OPPOSITE PAGE TOP Domino, or Keith the Beef, wreaking destruction. BOTTOM Raven's cunning honey-plan, to tempt this calf to feed from the bottle, worked a treat.

Violet and Sidney scrutinize the vet's handiwork following the calf's hernia operation.

Clive, armed with Reuben's metal detector, searches the calf suspected of swallowing my wedding ring.

We loaded everything back into the trailer and climbed back onto the quad bike.

No key.

'Oh, for Christ's sake, that bloody Sidney,' said Clive, looking at his son through narrowed eyes.

He lifted Sidney out of the trailer and tipped him upside down to check his wellies, then frisked him. It was no good asking Sidney. His verbal skills were limited. He could clearly say 'meat pie' if you asked him what he wanted to eat, but you couldn't have a sensible discussion about the whereabouts of the key. He would just open his eyes wide, blink, and give you one of his great big smiles.

We all combed the area looking for the key, a fingertip search trying to check everywhere that Sidney had wandered. But, busy with the shoot, I hadn't been tracking his movements carefully. He'd definitely been seen down by the rocks, and the other children had seen him near a boggy patch . . . The possibilities were endless.

Finally we gave up, left the bike and walked back to the farm-house, a considerable distance. The photographer was not happy, with all his gear to carry. Sidney wasn't happy, either: it was a long way for his little legs. Every so often he'd stop, look at me, and stretch out his arms, asking to be picked up. But, certain he was the culprit, we made him walk.

Back home I went through the filing cabinet looking for the bike manual, in the forlorn hope that with the documentation there might be a spare key. The search yielded only a tyre pressure gauge and a spanner for locking the wheel nuts. It was the weekend, so we couldn't ring Yamaha for a spare key, and when we did get through to them I knew it would take time and cost money. So our only hope was to take the metal detector up to the moor.

Clive, Reuben and Miles set off. They found all sorts of rubbish: tines off the haybob, bits of fence, but no key. Finally, Clive and Reuben discussed the possibility of hot-wiring the bike to get it back to the farmyard. That's when Clive peered down a small hole next to the steering column of the bike . . . and found the key, hooked up on the wires and sensors. It had only been inches from the ignition all the time.

I'm happy to say Sidney's fixation with keys has dwindled, and he has moved on to a much easier obsession to manage: manholes and gratings. Luckily, we don't have too many of these at Ravenseat and I suppose this is why they hold such a fascination for him. Recently Sidney was invited by friends to a day out at a wildlife park and animal sanctuary. He was not at all impressed by the animals (we've got plenty of wildlife at Ravenseat), but was fascinated by the drainage system. As my friend said afterwards, when I asked how the day had gone:

'It was great . . . grate after grate, we looked down 'em all!'

4

April

The bond between a yow and lamb is incredibly strong, some-times even surprising those of us who reckon we have seen it all. We try to gather the sheep down from the moor before their lambs arrive, but we don't want them in the lambing fields too early, or all our precious grass will be eaten before the lambs appear. It's a tight call, and inevitably a few yows will give birth at the moor.

Our sheep that graze on Birkdale Common are the ones that are most likely to run into trouble. In the 1960s, drainage chan-nels were dug in an attempt to improve the land by drying out the boggy, sodden ground. The ditches which crisscross our heaf have been eroded by water over the decades, and are quite deep in places. The heather grows along the sides, completely conceal-ing the ditches and making them treacherous places for young lambs. The grips (our name for the ditches) cause no problems for the yows, who have been brought up to know their own patch of the moor, and jump across the grips with ease. But if a newborn lamb slips in, the chances of rescue and survival are very slim.

One morning I was out foddering the yows on the common when I noticed that one of them looked as if she had lambed.

There were traces of blood on her tail, and I could see she had a full udder of milk. I scouted around but could not see her lamb. I waited until she had finished eating, in the hope that she would lead me to it. Sure enough, after I'd twiddled my thumbs for a little while, she made off away from the flock towards a large hollow known as the Punchbowl, but which we also call the Bermuda Triangle because of the number of mysterious disappearances of lambs in the area. I stalked her on foot, ducking down out of sight if she turned towards me, trying not to spook her.

She came to a halt on a lump of black, heathery ground and bleated loudly. I watched, expecting to see a lamb peering out of the vegetation, or hear him replying to his mother. After half an hour of her wandering about the spot I couldn't wait any longer and went to look for the lamb. There was nothing to be seen except a tangled bed of heather and seaves (rushes), and looking for a lamb amongst it was like searching for the proverbial needle in a haystack.

I don't like to leave a job half done, but I knew I couldn't sort this problem on my own so I went back to the farmhouse for reinforcements. I wrapped the children up warmly, and we all, including Clive, headed back up there. The yow was still lingering around the spot, and eyed us all suspiciously. I offered a prize for anyone who found the lamb, which set the children to work enthusiastically, but as time went on the sighs became louder, the faces longer. It was clear we would be going home empty-handed. Clive didn't seem surprised.

'She'll 'ave lost t'bugger. Bound to 'ave. It'll be dead somewhere.'

'What do yer wanna do?' I asked.

'We'll tek 'er yam an' I's gonna give 'er another 'un,' he said.

I reluctantly agreed that as her full udder of milk was going to be wasted, we should give up on her lamb and mother another one onto her. One of our neighbours lambs earlier than us, and he had mentioned he had some pet lambs in need of foster mothers. Back to the farm we went, this time for Bill the sheepdog, who helped us round our yow up and load her into the quad bike trailer.

'Shove 'er in t'stable an' let's be 'avin' some dinner,' said Clive.

We had a quick bite to eat and then Clive pulled on his coat. 'I'll away and git that lamb frae next door,' he said.

'Aye, I'll be out in a minute when I've put the dishes away,' I said.

He'd only been gone a couple of minutes when the kitchen door flew open.

'Yer did a crackin' job o' shuttin' t'stable door,' he said. 'T'owd bitch 'as gone.'

'Nivver bother getting t'lamb. I'll sort it now,' I said.

Clive stomped off into the barn. The children had lost interest in the whole saga by now so I jumped on the quad by myself and rode back up the road, certain that she would be heading onto the common. Sure enough, there she was, standing at the cattle grid. I knew I was supposed to catch her and return her to the farmyard, but something of her desperation struck a chord with me, and I just opened the gate and watched as she trotted back out along the track. I set off after her, keeping my distance, although this time she never looked back. It was only when we had travelled nearly a mile back onto the Punchbowl that she broke stride, coming to an abrupt halt and bleating loudly. I switched off the bike engine and sat quiet. She bleated again, and this time I heard the faintest of replies. I strode towards her and she never moved, only backing away from me when I was inches from her. Getting down on my knees, I parted the clump of heather directly in front of me and peered down into the dark, wet ditch.

There, about a metre down, was a lamb, almost indistinguishable from the muddy, peaty soil it was lying in. I lay face down on the ground and extended my arms down into the grip. My fingertips were just grazing him, but eventually I was able to grab just his lug, which made him bleat. It sounds brutal, but I lifted him by his ear until I could get a firm grip on his slippery body. Cold, wet and very dirty, he emerged squinting in the brightness of the weak spring sunshine. The yow, who had watched patiently, now stepped forward, and I swear I saw a look of sheer relief in her eyes.

'C'mon, mi lass, let's get yer both back yam,' I said. Holding the filthy lamb in my arms, I walked to the quad and trailer as the yow followed me obediently. I laid the lamb in some loose hay and she willingly hopped in beside him.

'Well I nivver,' said Clive as we drove back into the yard. 'It's not so oft tha' gets a live lamb out of a grip.'

Apart from being very hungry and dirty, no harm had come to the lamb, and he owed his life to his mother's stubborn refusal to give up on him.

There is, at the very start of April, a feeling of a lull before the storm, an awareness that at any moment we're going to be in the thick of it all, but we're waiting, waiting. Anticipation and excitement build: we know we need to do everything we can to make sure the number and quality of the lambs secure our future for the coming year.

The gimmer hoggs, 250 of them, will return to Ravenseat after their winter away. The farmers whose fields they have grazed are keen to wave goodbye to them, to let the grass grow before they let their cows out in the fields. Our hoggs should be fit, strong and well grown: that's what we've paid for. Well-wintered hoggs will go on to become big, solid, breeding yows.

As soon as they are back we must sort them out: it's vital they go back to the right heaf, as their inbred homing instinct is so strong that if they are in the wrong place they won't settle, and will stray in their urge to get back onto familiar turf. This is when we hornburn them, which means we stamp an indelible Ravenseat mark onto their horns, to show they come from here. It is one of the traditional ways of identifying horned sheep. Our marks are either the initials AC or CA, which may date back to the Campbell, Coates, Cleasby, or Alderson families, all of whom have farmed up here over the centuries. It might seem a little odd to be stamping the initials of somebody long gone upon the sheep, but it's all part of the heritage of the place. I also reckon that it can't be entirely coincidental that C and A also stand for Clive and Amanda . . .

To hornburn them we light a fire in a small empty metal oil-drum which sits on the wall of the sheep pens. The drum has holes cut into the side to let air circulate to feed the intensely hot fire within. When it is glowing red, our irons are laid in the embers. We have two irons for each set of initials, so that when one is being used the other is being heated. I suppose 'having other irons in the fire' is a commonly used phrase that really applies here. For in order to keep the job running smoothly the irons have to be constantly swapped to keep them searingly hot so they burn deep enough for the mark to be clear. The sheep don't feel it: horns have no nerves. The most they feel is a bit of discomfort at being held still. It's a smoky, hot job, but there is a strong sense of keeping a tradition alive, a link back to those previous generations who farmed up here.

Once lambing starts, we will eat (sometimes literally), breathe and talk sheep for the duration. Clive and I play a stupid game. He points to a sheep and says to me, 'D'you knaw who that is?'

'Mmmmm, nope . . .' I'll say, studying the woolly creature.

'Well thi should, because that's the one that . . .'

Then he reminds me about some incident or other with that particular sheep – it can be from years ago. It proves that Clive is a good stockman (or that he doesn't get out much). I know many of the sheep: the bad ones, the good ones, the old favourites. I know the wild ones who will put their heads down and refuse to move for the dog, and the wanderers who will turn up late at somebody else's pen miles off their patch. I also know the gluttons who will take a swipe at your legs with their horns and trip you up when you're feeding them. Despite the fact that the faces in the flock change year by year, we both ken them. There are small but distinct physical traits that are passed down from generation to generation within a flock. So rather than claiming that you know every sheep individually, kenning your sheep is about being able to recognize your own type from others of the same breed.

Raven has developed a good eye for sheep: at marking time when we are recording pedigrees she can look at a lamb and tell you who its father is.

'I bet that this is getten wi' Dazzler,' she'll say. 'Look at the white on 'is snout . . . An' that 'un must be a Battler, it's a gurt, lashy sort.'

Dazzler sires quite extreme lambs, with big heads, white lugs and fancy legs, all traits extremely desirable in Swaledales. At one time we had a tup called Stevie Wonder that got lambs with black patches in their wool, 'brooked uns'. His lambs would never be show winners but they were healthy and strong, so we did keep a few, one of which sports a big round black patch on her side, rather like a target, and is predictably called Spot.

We like as many of our sheep as possible to lamb outdoors, in the fields near the farm. The ground undulates, with bumps and hollows, there are beds of seaves and a few wooded copses, terrain in which the yows like to roam in search of their own private place to give birth. They like to find secluded spots, perhaps sheltering against a stone wall, shielded from the elements. It's natural and it's healthier too, as there is less chance of infection out there in the open air. Of course, it's not possible for all our sheep to lamb outside: for the very oldest and the shearlings (the first-time mothers) that have been scanned for twins, it would be folly to leave them to their own devices, so they must lamb in the barns.

As with everything up here, the weather dictates our lambing policy. We can't make plans as to how many yows will be in the barns and how many in the fields until we know what the conditions will be. We don't want it wet, cold or, worse still, snowy. But we also don't want it to be too warm. Warmth may bring a surge of new, green grass for them to eat, but it can also trigger an outbreak of a very nasty ailment called rattle belly (sometimes also known as watery mouth). It only affects lambs, and we easily recognize it from their excess thirst, drooling, sunken eyes, lethargy and a bloated belly that seems to rattle if gently shaken. In the very earliest stages it can be cured with a dose of antibiotics, but if it is left untreated it can be fatal.

Wet weather brings a different problem: joint ill. This infection is picked up through the newborn's wet navel and travels to one or more of the lamb's joints causing extreme pain and lameness. In dry weather the navel dries up very quickly after the birth, and the lambs born in the barns always have their navels dipped in iodine as a preventative treatment.

Changeable climate conditions can cause more direct problems. Heavy rainstorms can transform a trickling beck into a frothing, raging torrent that can wash away a lamb. A wet, wild, stormy day takes its toll on the yows, unsure whether to graze or just shelter with their backs up. Stress or extreme physical exertion can leave them susceptible to lambing sickness and staggers – both metabolic disorders, the first caused by a calcium deficiency, the other by a magnesium deficiency, and both with roughly the same clinical signs and outcome if not treated promptly. The affected yow loses her balance, staggers, as the name suggests, then goes down; her breathing becomes laboured; she slips into a coma and dies. A big syringe full of 60mls of calcium and minerals (which covers both ailments) injected under the skin or (if you have the nerve, or the situation is dire enough) into a vein, cures staggers easily – if you catch the problem soon enough, within a few hours the yow will make a seemingly miraculous recovery, and be back to grazing happily. But if you leave it too late, she'll certainly die. Vigilance is the key: shepherding is all about watching your flock.

One year we put a small flock of yows which were expecting single lambs into the Campbell Pastures, a very steep, rough pasture bisected by a deep ravine. We can't get in there on the quad bike, we have to go on foot; its very inaccessibility is why the yows like it, choosing to give birth away from the flock and us. But it's not easy if something goes wrong. I fed them daily, climbing over a stone stile and whistling to alert them that dinner was about to be served, then waiting patiently with Annas until the sheep came running, each day more of them bringing their healthy lambs. Fifty yows had been turned into the pasture, and because of the terrain it could take a while for them to turn up for food. If I was one or two short, I walked to the bottom of the pasture, crossed the beck, and peered across into the ravine from the other side of the valley. I used binoculars to spot them, and if they clearly had a newborn lamb then I knew it would be a day or two before the lamb was strong and agile enough to follow its mother back to the flock.

One day I noticed that one bigger, older lamb was obviously very hungry. As the yows put their heads down to eat the line of cake from the ground, she would nip in behind the row of woolly bottoms and, working along them, would try and suckle a mouthful here and there while the yows were distracted by the food. This was a problem, for while they were feeding I was not going to be able to work out which one was her mother.

'C'mon, Annas,' I said, 'we'll tek this lamb yam an' give 'er some milk.'

Annas, twenty months old, didn't reply, but she knew exactly what I meant, holding her little arms out to carry the lamb, well pleased with her new friend.

Back home, we handed the lamb over to Raven, who took her to the kitchen to mix up a bottle of replacement yow milk.

'Will she suck, Mam?' she asked. 'Or will I need to tube feed 'er?'

'Nae, she'll suck. Just fill 'er up an' we'll tek 'er back an' find 'er mother.'

She did sup all the milk and, freshly fortified, we all went back up to Campbell Pastures, and sure enough, now all the cake had been hoovered up, only one yow remained, frantically running back and forth bleating. I used the lamb to lure her into a corner of the field, where we surrounded and captured her. Raven examined her.

'Not a drop o' milk, nowt,' she said, shaking her head.

'Sorry, girl, you can't 'ave yer lamb back,' I said to the yow. 'You've done a good job, she's a good lamb, but she needs milk and you ain't got any.'

The yow was not happy, but that's the way it goes. We loaded her into the bike trailer, put a red stripe on her rump and put her through the moor gate to join the geld sheep. The lamb went back to the farm and into the pet lamb pen.

Annas was pleased to have her lamb back, and helped bottle-feed her every day. But one morning the lamb could not stand. It was really strange: she was still as bright as a button, but when we tried to get her on her feet she would dother and then collapse in a heap. It wasn't joint ill, because her joints weren't swollen. I wondered if it was because of a lack of colostrum at birth. Whatever it was, it didn't look good.

I gave her a shot of penicillin and separated her from all the other pet lambs because they can be rotters, jumping all over the weaker lambs. We didn't need a pen for her, as she was immobile, settled on a bed of straw, her legs folded neatly beneath her, at the top end of the barn where there are a few chairs we sit on to sup tea from tin cups and discuss the day's events. Every afternoon Annas would sit beside her in her straw nest and bottle-feed her, then curl up and doze with her head on the little lamb's woolly back. They looked so peaceful together, but I had my doubts as to whether the lamb was going to survive.

Things didn't improve. The lamb still had bright eyes and fed enthusiastically, but she made no effort to stand, and when we tried to help her she collapsed. I pored over *The Veterinary Book for Sheep Farmers* – it's always a sign of desperation when this dusty tome comes off the shelf and onto the kitchen table. I could find nothing that fitted the description of her symptoms, and Clive broached the possibility of putting her down.

'Poor lal' bugger canna' stand up,' he said.

'I know, but . . .' I knew he was probably right, but I wanted to soldier on a bit longer.

Every day we'd scrape away her nest of wet, dirty straw and replace it with clean, dry bedding, always hoping that when we opened the barn door in the morning we'd see some signs of recovery. For ten days it was dispiritingly disappointing, and even I was beginning to wonder how much longer we could persist. Then, one morning, the little lamb was not in her nest. She had moved a few yards. She was still lying down but nevertheless it was progress. There was a light at the end of the tunnel and it was the start of a slow recovery process, lasting a few weeks. She began following Annas for a few stiff steps around the barn, progressing to short rambles around the farmyard, always accompanied by Annas. It was a joyful day when we put her back with the other pet lambs and watched her running laps around the pen. We don't always get happy endings, but it's jolly nice when we do. To this day, we have no idea what caused her condition.

The sheep we keep inside to lamb have to be watched day and night, and we take turns getting up through the night to check on them. It's not natural for sheep to lamb in an enclosed space, and for every problem it solves, it creates others. It's safe, it's dry and it's out of the weather, with help close at hand, but infection is more prevalent lambing inside. After a birth inside we put the yow and her newborn into an individual pen as soon as possible, to avoid any mismothering. When the shed is full of expectant yows it is easy for a lamb to get away from its mother, especially if it is one of twins. The mother will be so busy with her second arrival that the first one may wander off. Once the lamb is separated it can be claimed by another yow, who may or may not have already lambed herself. With their maternal hormones raging, some yows are hell-bent on stealing newborn lambs – and once it has been stolen the real mother may reject it when they are reunited, because it smells of another yow.

It has been known for half-asleep Clive or me to pen a yow with a newborn lamb in the wee small hours, noting that she has been scanned to have twins. We go back to bed leaving nature to take its course, only to find the next morning that the pen wasn't properly secured and now she has three fine lambs . . .

'This dunt add up. I smell a rat . . .' Clive will say.

The mystery is solved when another yow who scanned for twins trots past with just one lamb.

One year, after a gruelling month of lambing through the night, we only had about twenty left to lamb and Clive said, 'T'ell wi' 'em, we're not getting up to 'em tonight. They'll be grand.'

I didn't argue. But the next morning six of the yows had lambed, each with twins, and they were all muddled up. Two of them had four lambs each, two were pretending they didn't have any offspring at all and the last two were beating up any lamb that came near them. It was a mess, and it took us an hour of juggling to end up with three sets of twins, three singles, and three pet lambs that we could have done without. That night, we set the alarm for two a.m.

Life becomes a long round of feeding and watering and cleaning out pens. We make sure all lambs are suckled and full, the yow's udder is checked, and if the weather is agreeable they go out to the fields, loaded into the bike trailer, which is specially adapted to transport three yows and their lambs.

The children look forward to lambing time; the first lamb to be born at Ravenseat is subjected to some serious stroking, petting and cuddling, and the novelty does not really wear off. For Clive and me, too, there is always magic in seeing new life coming into the world, and even after sleepless nights and disheartening weather, there's nothing like seeing a band of lambs racing around the fields and frolicking in the late-evening sunshine.

There are occasional casualties: giving birth is a risky time for any animal. If a yow loses a single lamb, a very good way to get the mother to adopt another lamb is to skin the dead one, and then the chosen foster lamb will be dressed in the new woolly overcoat. He will be introduced very carefully to his adoptive mother, and if all goes to plan he will try to suckle while she sniffs his tail. Reassured by the smell of her own lamb, she will accept him. There is no hard and fast rule as to how long the skin stays in place. Sometimes, with a particularly motherly yow who accepts the lamb very quickly, the skin need only remain in place for about an hour or so. For other more suspicious yows, the skin may stay on the lamb for a day. That really is the limit, for after this time the lamb will become quite whiffy and I defy any sheep to love something that smells so bad. The whole process may sound barbaric but the reality is that you have nothing to lose – the worst that can happen is that the yow rejects the lamb. If it works, then you have a happy lamb matched with a happy mother who has an udder full of milk for him. It's a traditional and very effective method and the children are fascinated by the whole process, watching with serious faces and occasionally asking questions.

'What's that bit?' Edith will say, pointing to a testicle.

The children are quick learners, involved with the sheep from a very early age; thus when they get older they can be entrusted with important jobs. They are thrilled when they've helped a yow with a difficult birth, perhaps even saving the lamb. They work as a team, but it's not exactly a well-oiled machine: there may be a bit of arguing about who does what. When they tell me later what they did, there may be a little bit of embellishment, but they take great pride in being able to make decisions and take matters into their own hands.

You can always tell what we've been doing at this time of year, as our hands are stained yellow from the iodine navel spray we use. We look like a family of very heavy smokers with nicotine-stained fingers. One year when I was expecting, I was quizzed by a doctor about my involvement with lambing while pregnant. I flatly denied being anywhere near the lambing shed, but I could see she was looking hard at my yellow hands and the blue foot-rot spray ingrained in my nails . . .

Triplets are rare in our flock, but one year we did have more sets than usual. We kept the five yows scanned for triplets separately in a stable, with extra rations on the menu. Raven took care of them and was very proud of her triplet yows: they really were in tip-top condition when lambing time came. Sure enough, they all gave birth to good, strong lambs with no problems, until she got to the last one.

'Mi last triplet yow 'as come a' lambin',' she said. 'I'll move 'er into t'hospital when she's lambed 'cos I'm gonna 'ave to 'elp 'er feed 'er lambs.'

She stood quietly by the stable door, taking heed of what Clive and I have always preached: 'Let nature tek its course, don't interfere unless yer really 'ave to. The best tool ye 'ave for lambing is yer eyes.'

The old yow laboured, then gently licked the faces of each of her lambs as they took their first breaths. There was no time to waste: the sooner the yow and her triplets were in the hospital pens the better, and then Rav could top up the lambs with colostrum stolen from yows who had single lambs. She managed to pick all three up, slippery with birth fluids, and headed to the hospital with them, walking backwards, back bent with the lambs held in front of her so that they were never out of the yow's sight. They proceeded along the corridor beside the pens, and all the while Raven imitated the noise of a bleating newborn in order to keep the yow's attention. It was only when Raven had the yow and her three lambs penned, and was tying the hurdle with string, that she noticed a tiny lamb lying in the corridor just a few yards away. It looked lifeless, with fragile matchstick legs and the smallest triangulated head imaginable, with a blue-tinged tongue lolling out. It lay, quite still, in a puddle of birth fluids.

'Never say die' is one of our mottos, and Raven didn't. She grabbed a handful of straw and started to rub the little creature vigorously. In a last-ditch attempt to revive the tiny mite, she blew gently through her cupped hands into the lamb's mouth. There was a faint gasp, a splutter, and a big smile from Raven.

Quads! Our first EVER set of quads.

The yow had enough to contend with looking after triplets; she was eventually turned out to the fields with two of her lambs, after a suitable foster mother was found for one. The fourth, a gimmer, was wrapped in a tea towel and shipped into the intensive care unit (the black range oven). She was given a little fresh, warm colostrum from a syringe, then Raven decided to christen her.

'She's Minty the micro lamb. Isn't she sweet?'

Alec had his doubts as to whether this was a worthwhile exercise, and as usual he didn't mince his words: 'Nivver 'ave I seen owt so small. It'll die, sure as eggs is eggs.'

'We'll see,' said Raven.

Clive was in agreement with Alec, but couched his words less bluntly. 'Some things just aren't meant to be, Raven. Don't be too upset if she dies.'

I thought back to Reuben's birth, remembering how premature and tiny he was, the problems we had, and how sometimes the tiniest babies can be real fighters, beating all the odds.

Minty didn't die that first night, sleeping soundly in the oven with me and Raven taking it in turns to feed her little and often. Within days she had graduated to living in a washing basket in the airing cupboard. She found her feet and she found her voice: a high-pitched bleat that would wake Raven from the deepest sleep to feed her. She spent three weeks living in the house. We often have lambs in the oven or by the fireside overnight, but for one to remain inside for so long was unheard of.

Every so often Clive would say he needed a lamb to mother on to a yow whose own had not survived, or who was fit enough to take an extra one.

'What about Minty?' I suggested.

'Nay, to hell, I's not putting that lal' fart on a yow, it'll nivver be strang enough an' it won't know 'ow to sook a yow,' he said.

When lambing was nearly over, with just a handful of yows left to lamb, we had a big strong yow with a decent udder full of milk who was scanned to have a single lamb. Clive was away in the fields when she went into labour, so Raven and I hatched a plan. We'd try to adopt Minty on to her. The worst that could happen was that she'd reject the tiny lamb.

Clive's not gonna like this, I thought.

'Dad's really not gonna like this,' said Raven, echoing out loud my sentiments.

We knew as soon as we caught sight of its feet that the yow's own lamb was huge. I wasn't confident about little Minty's future. We followed the usual wet adoption process, laying a plastic feed bag that had been split down the seam under the yow. Minty was hobbled, and hidden close by in a hay rack. Tying the legs of a foster lamb is essential if the yow's suspicions are not to be roused, and it also prevents the foster lamb taking all the colostrum before the yow's own lamb has had time to get to her feet and suck.

Raven held the sheep steady while I delivered her lamb, catching all her birth fluids on the plastic bag. Minty was then 'rebirthed', doused in the fluids and rubbed alongside the newly delivered lamb. When Minty was sopping wet I put her next to the yow's head, introducing the impostor lamb to her adoptive mother. It was a critical moment: would the yow accept her? To our relief, there was never any doubt in the yow's mind that this was her lamb. She licked the little thing, nuzzled her, bleated quietly. Minty lapped up the attention. When we were sure they had bonded, we gave the yow her real lamb.

Raven and I left them all together for about a quarter of an hour, to get to know one another. When we returned, we untied Minty's legs. This was another critical moment. Would she be able to suck? It's a natural reflex in a newborn lamb, but Minty had been bottle-fed for three weeks, and that can diminish the instinct. When she stood, she could still barely reach the teat, but she got herself latched on and, standing alongside her enormous 'twin' sister, she guzzled away.

We kept them in the stable for weeks, until we reckoned Minty was strong enough to keep up with her mother and sister out in the garth. Clive would shake his head whenever he saw the little family unit: one huge lamb, one tiny one. But six months later Raven and I knew we were vindicated and forgiven. We were speaning (weaning) the lambs when a familiar face came into the pens. Minty had thrived, and was now a similar size to the other lambs.

'Yer know what, I's thinkin' tha' this un's a keeper,' Clive said, as she came down the sorting race. 'She's a nice sort, good quality, I like 'er.'

This was the highest praise. Raven was tremendously pleased, and went off to the house, returning with a special one-off ear tag in purple and brown, so that Minty would always stand out from the rest of the flock with their green and yellow tags.

The last job before night falls is to go round the fields, checking on the yows and any newborn lambs. Then they are on their own from darkening to first light, about 5.30 a.m., when we set off on our rounds again, but this time with a bag of cake and bales of hay.

For some reason, rain makes sheep lamb: lambing in dry weather is always slower. Even the ones in the sheds lamb faster when rain is drumming on the roof. You know when you do your morning patrol of the field after a wet night that there will be a bumper crop of lambs.

Within minutes of its birth, a lamb takes its first, faltering steps, and within a couple of days it will be able to outrun the fittest of shepherds. This is where the traditional shepherd's crook comes in handy as an extension of the arm, to catch a lamb or yow that needs attention. Cunning, stealth and a certain amount of luck are needed to catch a lamb, creeping up behind him before he realizes that you are there. He can accelerate from nought to full speed much faster than you can, so the best chance is the first attempt. It can be very frustrating. There is an old saying that if a shepherd is not fit with a stick, he should leave it at home: in other words, if he is likely to lose his temper and hurl the stick at the escaping lamb or yow, he should leave the stick hanging over the mantelpiece.

Clive and I have always differed in opinion as to which type of crook is the best. When I worked as a contract shepherdess I always used a leg crook, the traditional and instantly recognizable shaped stick, with which you could catch a yow by its back leg. My argument with Clive was that most of the time when you need to catch a sheep you are sneaking up behind it, so its leg will be the nearest part to you.

Clive rates the neck crook, the stick with a wider end which you can slip around a yow's neck, preventing her from moving forwards. In his opinion the chance of being able to keep hold of the neck stick while moving forward and grabbing the yow is infinitely better than with the leg crook. He may be right, but both types have their advantages and drawbacks. Nine times out of ten, you don't have your crook to hand when you need it anyway.

Catching the lamb or yow usually leaves you out of breath, laid out in the grass or heather, hanging onto the yow's horn or with a squirming lamb in your arms. Therefore it's important to have everything you need to hand so that you don't have to drag your patient to the medicine. What I can't fit in my pockets, I carry in a well-worn satchel. In here are the lambing-time essentials: an aerosol of blue antiseptic spray, a small empty screw-top bottle and teat, a feeding tube, bottles of penicillin and calcium and mixed minerals, and syringes to inject them with. Also lurking in a small zipped compartment might be a little bar of chocolate, purely for medicinal purposes of course, for when the shepherdess's spirits are at a low ebb. Being able to treat ailing sheep immediately can mean the difference between life and death.

One of our neighbours remembers shepherding on the moor many years ago and finding a yow laid out on the heather, her breathing shallow and her eyes glazed – clearly staggers. Fortunately he had calcium with him, but unfortunately syringes in those days were made of glass, and his had shattered in his bag. Rather than leave the yow to die he decided to use his penknife to make an incision in the skin on her inner thigh, using his fingers to make a pouch between the skin and flesh into which he poured a dose of the liquid calcium. He gently massaged it until it was absorbed. Reckoning he could do no more, he left her to attend to the rest of the flock. When he came back later she was on her feet, nursing her lamb. His improvisation had worked.

It is frustrating to find yourself without a knife in your

pocket: it is vital for cutting twine, or opening a bag of feed, or even for giving sheep pedicures. Knives get lost on a regular basis, but Reuben has a knack of finding them when Clive or I have only the vaguest idea of where we have lost them.

'I was in t'Close Hills when I last 'ad mi knife on mi,' Clive says.

Reuben will be back with the prized knife before the day is out. He only once failed, and that particular knife had a green handle – clearly not designed for the farming fraternity, who drop them into the grass. It was desperation due to the number of objects we lose while going about our daily business that drove us to buy a metal detector. Although it was primarily intended for work purposes, it wasn't long before Reuben was setting off on a mission to find buried treasure – although his definition of treasure does not always tally with mine.

'Mam. Look what I've found!' he'll say, excitedly fishing around in his pockets. 'A blade off a cutter bar.' He proudly holds out a rusty lump of metal.

We once had a metal-detecting enthusiast visit Ravenseat, and I hoped that he would give Reuben some tips on how to go about it. He was a man of few words, but after bribing him with tea and home-made cakes, he agreed to walk around the farm with Reuben and his detector, showing him the ropes. A couple of hours later they returned, as pleased as punch to show me what they thought was a medieval groat.

'A nice example,' said our visitor.

And a small rusty bell.

'Perhaps off the harness of a packhorse,' he said, rubbing it on his shirtsleeve. There were also a couple of pound coins that must have been dropped by the cream-tea visitors.

'Ah, well, nivver mind. I don't suppose yer ever find owt of any real value,' I said.

'Well, actually . . .'

He told us about an afternoon of fruitless detecting in a field at Kirkby Stephen in 2010. He and his son had been given permission by a farmer to detect on his land, but apart from the occasional piece of rubbish, they had found nothing. It was getting late in the day and they were considering packing up when the detector bleeped, alerting them to an object hidden thirty centimetres below the surface. Putting a spade into the soil, they levered up what looked like the rim of a Victorian-style brass coal scuttle. It was only when they began brushing away some of the loose soil and saw a metal griffin that they knew they had found something incredibly important, and that their lives were going to change forever. It was the Crosby Garrett helmet, a copper alloy Roman cavalry helmet from the late second or third century AD, and it later sold at Christie's for £2.3 million.

I was surprised that the man who had tramped around Ravenseat with Reuben for an afternoon and had enthused about a rusty groat had once found something so special – the ultimate find that could never be bettered. He explained that his enthusiasm for detecting had not waned, and the only difference was that he now had a top-spec detector, the best that money could buy.

Some of our older yows, the ones we affectionately call the crusties, can become so protective of their lambs that they have been known to attack us when we've ventured too close. Everyone assumes it's the male of the species that should be treated with caution, but when it comes to aggression there's nothing like an angry yow. We had one in the pens for a few days because her lamb was dopey and slow to feed; we had to be very careful when filling her water bucket as she would back up to the rear of the pen and try to butt anyone who came too close.

One of our old ladies, who we'd shown very successfully when in her prime, gave birth to two good-looking specimens, certainly with show potential: a gimmer lamb and a tup lamb. Clive brought her in from the fields and back to the farm, saying: 'She's not feeding 'em sa' weel. She's got milk on both sides, but mebbe not enough to fill 'em both.'

'We'll mebbe have to take yan off her, she's too owd to look after 'em both. I'm sure we'll find a foster mother for t'other,' I said.

There was no need to make an immediate decision, so I gave her extra feed rations and kept her in a pen, to let her try to keep them both. The old crusties are all good mothers, very attentive; it seemed a shame to take one lamb away. She stayed in the pens for a couple of days with a full hay rack and plenty of food, and when the sun was shining we put her in the garden for a nibble of grass. She seemed to be managing to feed her twins, and we were hopeful that she'd turned the corner and that she'd now make enough milk to feed them both. We turned her and the lambs out into the fields, thinking the job was done.

But the following morning when Violet and I went into the field we found both lambs as flat as kippers, their little bellies empty. It seemed clear that she just couldn't make enough milk to sustain them both. I brought them all back to the garden for Clive to decide what to do.

'Nah, we'll tek the tup off, mek a note of who 'is mother is, give 'im some bottle an' put 'im in wi' t'pet lambs.'

Clive wasn't best pleased: it was a good-looking tup, and they always thrive better with their own mothers. But I had to agree. 'No, yer right, she can't feed two. I'll put him in t'pen . . .'

The tup lamb was a big, strong fellow, well marked and a really good specimen of a Swaledale. He stood out head and shoulders above the other pet lambs, literally because of his size, but also because he was such a fine lamb. We were hopeful that he wouldn't be in there for too long, as he'd be first in line for a good foster mother. The yow and her remaining gimmer lamb were taken to Close Hills pastures, where the yows with single lambs grazed.

The next day was very busy, with lots of lambs born, and by the time evening came we were ready to sit down. The children had been playing in the yard all day and were black bright (very dirty) so I'd been overseeing their ablutions, and finally had them all in their beds. Clive and I settled down on the sofa for a well-earned respite from our labours. It was darkening, the fire was blazing, our coats hung above the fireplace and our wellies were ready by the door to pull on over our pyjamas at some ungodly hour, to check on the sheep in the barns.

We had just got comfortable on the sofa when we were roused by the sound of falling stones, and a very loud bleating noise.

'What the bloody 'ell is that?' said Clive. 'Is there no peace for the wicked?'

'Well, I'm no expert, but it sounded very like a yow to me . . . C'mon, let's ga an' see what's 'appening.'

Standing in the garden was the yow we'd separated from her tup lamb the previous day, with her gimmer lamb at her side. She wasn't happy, and she was being very vocal about it. Quite how she had made the epic journey from Close Hills we still don't know, but along the route she must have gone through at least three gates. Or, more likely, she'd scaled the walls just as she had done to get into the garden, sending all the topstones flying and making a sheep-sized gap.

This was not good news for Clive. If there's one thing he hates, it's ratchin' sheep clambering over walls or squeezing through fences, teaching their offspring and others to follow in their wake. Just a tiny bit of wool caught on a wall-top wire and blowing in the breeze signals an escape route to other members of the flock.

'Bloody thing,' he muttered, as the yow looked at him and *blaaaaarghed* loudly.

'I ain't no sheep whisperer or owt, but I reckon she's come back for 'er lamb. This is t'last place she saw 'im afore I took 'im away,' I said.

'Well, she best 'ave 'im back then,' Clive said, setting off towards the pet lamb pens.

He came back a minute or two later carrying the tup lamb, who, upon hearing his mother's call, struggled to break free from Clive's arms. Clive set him down by the garden gate: he ran straight back to her and immediately dived underneath to suckle, while she turned her head and tenderly nibbled his tail top.

We both leaned against the gate in the half light, watching the yow, who was now oblivious to our presence. I smiled. Even Clive was smiling as he said, 'She'll 'ave to keep 'em both now, we'll just 'ave to keep topping up t'lambs wi' t'bottle.'

'Yer must be gettin' soft in yer owd age,' I said.

We couldn't have kept them apart after that. It had taken a monumental effort on her part to get back to her lamb, and if there was ever proof needed that sheep are not stupid, this was it. Every day, either Reuben or I would take a bottle into the field for the lambs. Eventually, they grew bigger and began to graze a little themselves. That, together with our top-ups and her milk supply, meant there was enough food to sustain them both. It's possible that the lambs would have grown bigger, and perhaps stronger, if we'd taken one away, but sometimes our hearts rule our heads, and I don't reckon there's much wrong with that.

Another problem that we have to deal with is overproduction of milk. It may seem strange, but a lamb can literally starve to death because the yow's teats are so engorged with milk that they are too big for it to suckle. This can easily be solved, but needs to be tackled quickly, because if the lamb misses out on suckling within the first few hours of his life he will soon weaken. The colostrum, produced at the very beginning, contains the vital fats, nutrients and antibodies needed to kickstart gut function and the immune system, and a lamb will go down fast without it. By instinct, as soon as he finds his feet a lamb will search for the teat. As always, close observation is key: milking the yow, suckling the lamb and getting the teat down to a manageable size. The extra milk comes in very handy for the lambs whose mothers are not producing enough, or the ones who are orphaned.

All the children, from Violet upwards, can milk a yow, and there's always an argument about who is going to do it. Filling little pots with milk is a satisfying job, and they all know to pop them in the fridge or freezer ready for the next hungry lamb.

I recently had a lot of hungry mouths to feed, but of the human variety. I decided to make a custard tart with our own lovely fresh eggs, milk, and a dusting of nutmeg. It went down very well – anything that isn't a jacket potato goes down well during lambing. It was only when it had all been polished off and Annas was actually licking her plate clean that I told them all that the milk was not from a cow, but from our own sheep. I hadn't told them for fear of putting them off but in fact it had exactly the opposite effect, and they were soon demanding that I bake another one.

When I was a jobbing farm worker I was once served a particularly unpleasant dessert that they said was junket. Unlike proper junket – made with milk and rennet, and flavoured with rum or clotted cream – this was made with cow colostrum, baked gently until almost set. It looked like custard, but by God it didn't taste anything like the Bird's stuff that I was brought up on. Clive was presented with the same thing when he was a hired lad on a farm, served by its proper name of Beastings Pudding. He remembers it as yellow slop, complete with bloody streaks . . .

You can tell when a lamb isn't getting enough milk from its mother. You get attuned to it: they just don't look as round and full as they should. If in doubt you can pick them up and feel the tummy. They don't need a great deal, just a constant supply, little and often. It's much harder to tell from the look of a yow if she is too thin, not thriving. Their woolly fleeces hide a multitude of sins, so you need to get your hands on her back and feel her backbone, to check for protruding bones. Just as too thin is bad, you also don't want her to be rolling in fat. A happy medium is best, although every animal is an individual and they differ in physique, character and mannerisms, just as people do.

With the shearlings, the yows who are lambing for the first time, we watch carefully. Many will have no problem at all adapting to their new role as mothers, but a few find the whole experience bewildering. They seem not to understand why they are being followed around by one, or possibly two, bleating, woolly little nuisances. If they don't have any maternal instinct we have to show them, encourage them. It can be hard work persuading a yow to love her lamb, but it's worth the effort, because we don't want to add to our pet lambs. We put the shearlings in a pen, where they can't abandon their offspring, and gently persuade them to relax as the lamb suckles.

There are days when we congratulate ourselves because we get through without putting any lambs into the pet lamb pen, and then the next day we'll have a rush of them. Our pet lambs do well: we have an expensive automatic feeder which mixes warm milk for the pet lambs, allowing them to suckle when they want, in as close a way as possible to the real thing. This gives them the very best start in life, if rearing naturally on a yow isn't possible. It is an expensive business though, buying the milk powder and feed. But it's all part of lambing time.

Easter usually falls in the middle of lambing, and for us the celebrations are a non-event. I do make sure I have Easter eggs on hand, hidden away for the children, and on Easter Sunday they find them in the henhouse, so the little ones believe that the hens have laid chocolate eggs. But apart from that, there's no special celebration. It's all hands on deck at lambing time, and their school Easter holidays are usually spent helping out among the flock. One year recently was unseasonably hot, but there was a breeze that took the edge off the heat, and fooled me into not noticing that the children were catching the sun. I was beside myself when I stripped them off for their showers and realized that the fairer-skinned ones had bright red ears and noses. Edith and Miles have olive skin, and over the summer slowly turn a lovely nut-brown colour. Sidney and Violet are not so lucky.

'Oh my gawd,' I shouted to Clive. 'What am I gonna do? They're bright red.'

'Don't worry, it'll calm down . . . an' you need to calm down an' all.'

There is nothing less likely to calm me down than being told, 'Calm down.'

'Are you kiddin', Clive? They've gotta go back to school soon. Having a sunburnt kid is on a par with carrying a class A drug nowadays. Unforgivable. I'm gonna be vilified,' I said, slapping on the skin cream.

As it happened, the weirdly unpredictable British weather meant that before they went back to school we'd had blizzards to contend with, and I'd had to dig the balaclavas and gloves out of the back of the cupboard. A chill wind blew, snow fell thick, and the sunburn quickly faded.

So April comes and goes in a wave of elation, sometimes disappointment, and worry. We are left emotionally and physically drained, an indescribable weariness demanding that we momentarily sit back, rest and take a little time to gather our thoughts before we gather the yows and lambs into the pens and begin marking up and turning away.

5

May

It was in May that Keith the Beef showed his true colours, on the day we turned the cows out. Having spent the winter cooped up in the cow shed, we expected some high jinks from all of them when their hooves touched the grass; and sure enough, they set off at a fair pace up the field, their udders swinging from side to side and a few enthusiastic kicks and bucks thrown in for good measure. Keith ambled out: he never did anything at great speed, but he kept a close eye on his harem as they frolicked in the distance. They went out onto the moor bottom, with a round feeder filled with haylage for them. After a few hours of exuberant capering about, the herd settled round the feeder, with Keith taking on the role of babysitter, looking after the young calves. But he was soon bored, and that's when he began his trail of destruction. He'd behaved well enough inside (the white bit of the domino), but we soon found out that there was another side to him (the black bit). He'd stroll along to the nearest drystone wall, reverse up to it and give his rump a good old scratch along it. No wall can stand a ton of muscled beef pushing against it, and it was not long before there was a loud crash as the wall tumbled.

Neither were the few wire fences we had any match for him.

He just leaned his great weight against them, his leathery hide untroubled by the barbed wire, and chomped away at the meadow grass on the other side. The fence posts leaned inwards, the wire was stretched to its limit, and before long the fence gave in to the pressure, twanged and sagged to the ground. Then he stepped over it into our precious hay meadows, leaving just a few tufts of red hair on the barbs – and was followed by the rest of the herd, who joined in the feast.

It was only when Keith broke into our new tree plantation, smashing the saplings and their protective tubes, that Clive saw red. He set off after Keith. 'I'll warm 'is fat arse,' he growled, striding off purposefully, wielding a pitchfork and following the trail of destruction.

It was one of those comedy moments: Clive running at Keith, threatening him with all manner of things and swearing. A nonchalant, thoroughly unflustered Keith lumbered away towards Clive's masterpiece of fencing, where he had actually re-routed the river to sink a huge strainer post into the solid stone river bed, concreting it in.

'Rock solid, that'll be forever,' he'd said at the time, admiring his own engineering. 'It'd tek a bulldozer to shift that.'

Well, he'd reckoned without Keith, who ambled through the fence without pause, leaving the 'rock solid' upright straining post leaning at a lopsided angle, a bit like the Tower of Pisa.

That was the final straw. He went back to his owner after that.

For the calves, born during the winter months in the confines of the barn, turnout day is their first taste of freedom and the great outdoors. Last year our three Beef Shorthorn calves spent their first few days lying in the loose hay round the ring feeder, never going far, just waiting for their mothers to return from grazing the sweeter grass further up the valley.

As they gained confidence in their new surroundings, they began to venture further and explore new territory. Every day we walked through the sheep pens, up the stone track to the stone stell (shelter) at the moor bottom, and carried out a head count to make sure they were all present and correct: four cows with three calves at foot, two heifers and one bull. Within days they were losing the winter-time accumulation of muck 'buttons' which hung from their bellies, and their red roan coats stood out from the green and brown moorland, making them easily visible at a distance.

They had only been out for a couple of weeks when one morning we noticed that one of the calves, Gloria, was not with the others. Her mother Eve did not seem perturbed, and watched disinterestedly as we conducted a search of the nearby seave beds, the most likely place for a small calf to hide out. When we could not find her, we crossed the river to look at the back of the moor wall: and there we saw her, in the beck, on a rock covered with slippery algae, just below the first drop of the Jenny Whalley waterfall. Gloria was standing rooted to the spot, water lapping around her hocks and a look of terror in her eyes.

'What the 'ell is she doin' in there?' Clive said.

'Who knows?' I said. 'Mebbe she slipped on the green slime on the river bottom when crossin' with 'er mother. I wouldn't 'ave thought she'd 'ave voluntarily waded out into t'water an' clambered down t'waterfall . . .'

We struggled down the steep banking, waded carefully into the river so that we didn't lose our footing, and edged slowly towards Gloria.

'Mi wellies are fillin' wi water,' Clive complained.

Gloria saw us approaching and decided she wasn't going to hang around any longer. She shot forward, skidding on the algae, and leapt upwards, landing on her knees, and scrabbling away with her back hooves. Her head was being doused by the waterfall, but she finally managed to get a foothold on the bank and scramble up to firm ground. She stood for a moment, dripping wet, looking forlorn, then shook her head and galloped across to the feeder where the herd of cows was nonchalantly chomping away, unaware of the drama. As Clive and I squelched our way back to the farmhouse we looked across at Gloria, now happily suckling from her mother Eve.

We worked for the rest of the day in the sheep pens, tagging and marking the lambs, getting them ready to be turned away to the moor. As the day wore on the weather turned, the wind picked up, the sky became leaden and we knew a storm was brewing. We didn't worry about the well-being of the cattle: Beef Shorthorns are a hardy breed, and can cope with any weather Ravenseat throws at them. As long as their feeder was full of haylage and their bellies were filled, they would weather the storm.

It was a bad night, the rain was beating against the windows, and in the morning when I peered out of the bedroom window there was a roaring flood. We don't fear floods: although we are surrounded by water we are at the top of a hill, which means we, and the farm, stay safe. But floods cause us logistical problems. We can't cross the river, but we can still use the packhorse bridge. Feeding the sheep can endanger them: they are so keen to get their heads into the feed bag that they will attempt to leap a raging beck, risking being swept downstream. Although sheep can swim, the weight of their own sodden fleece makes it difficult for them to stay afloat. On a really bad day the sensible option is sometimes to leave them be and wait for the water to subside before venturing out with food.

That's what we decided to do. Clive and I busied ourselves around the yard, while the children splashed in the puddles and played at filling buckets from the broken gutter on the barn.

'C'mon,' I said, 'let's ga an' see what the coos are up to.'

'I'm gonna tek the lal' uns in and get t'kettle on,' said Clive.

The older children, Raven, Reuben and Miles, came with me up the stone track. My woolly hat was wringing wet, and I kept my hands thrust down into my pockets. Looking up at the sky-line, I could see a plume of spray above the High Force waterfall: the usual trickle of water, barely visible at this distance, had become a torrent of white water, cascading and tumbling over the rocks, bubbling and foaming as it carved out a wider water-course down the valley.

The cows, although sodden, were lying in the shelter of the stell, chewing their cud placidly, glancing at us without interest as we counted. One short, again. And again it was Gloria.

'C'mon, hup mi lasses,' I shouted, hoping that Gloria was concealed amongst them. They reluctantly stood. No, I was def-initely a calf short. I scouted round, looking in the clumps of seaves. No sign. Why would she be away from the herd on such a horrible day? I remembered the previous day's events, with foreboding.

'Stay here an' play spot the calf,' I said to the children as they stood dejectedly, looking this way and that for our missing calf. I walked down to the beck, where she had been marooned the previous day. But what a difference: where Clive and I had waded in, the roaring, turbulent waters now ran deep, brown and dangerous. My heart sank. If Gloria had gone back to the river, she would have been swept to her death.

'You ga yam,' I yelled to the children, 'Tell thi father what I's on with.'

I didn't dare risk crossing the river, so instead walked the riverbank back to the farmhouse. I knew that if she had been washed this far downstream then realistically she would be dead, and I was effectively looking for a corpse.

When I arrived back at the farmhouse Clive was already getting coated up, and he set out to follow the river downstream to the boundary of our land. He found nothing. Gloria seemed to have vanished without trace, but the mystery was that Eve, her mother, was once again completely unperturbed. We discussed it and came to the conclusion that Eve might have hidden her calf, and possibly knew exactly where Gloria was. The answer was to put Eve under surveillance, and Raven willingly volunteered to do the undercover job. She stationed herself in the tractor, which was parked behind the sheep pen wall, with a pair of binoculars and a book to relieve the tedium. She didn't stick it for long, and it soon fell to Miles to take her place.

Miles is a thinker – quiet, content in his own company – and he was happy to keep Eve under observation. It was he who, after a few hours, spotted her leaving the others and setting off at a slow, determined pace up the moor. She wasn't mooching or stopping to graze; she was on a mission, and it was Miles's job to see where she was going. He stalked her, keeping a safe distance, and ducking out of sight when he thought the cow could see him. She didn't cross the river, but travelled up the valley bottom to a point where the river forked. Several yards away stood a half-collapsed Nissen hut, where hay had been stored in years gone by. Eve quietly lowed, and a very dry, contented-looking Gloria emerged from the remains of the hut.

Cows are wise, often more so than we give them credit for. After Gloria's escapade in the river, Eve had been determined to keep her calf safe. We were all very pleased when Miles delivered the news back to the farmhouse.

We use our cows as a barometer, to tell us when better weather is in store. Not by the old wives' tale about cows lying down when it's going to rain: they lie down to chew the cud. But if our cows leave the shelter of the lower slopes and make their way to the moor tops, then it is a reliable sign of settled, dry weather. There is nothing more pleasant on a lovely evening than seeing them quietly grazing the heights, amongst the heather, oblivious to the occasional haunting trill of the curlew and the staccato cackle of the grouse.

At the very top of the moor is Red Mea Tarn, 1,800 feet above sea level. It is a bleak, lonely place. In winter its inky black waters, hemmed in by exposed peat haggs, ripple in the icy wind that drives across the desolate expanse of open moor until the temperature plummets and it freezes over. In late spring it wakes from its winter slumbers with the arrival of black-headed gulls, returning to one of their regular breeding colonies, where they lay their eggs on the shoreline.

The children were always a little wary of the tarn, shivering at the thought of its unfathomable depths. After the initial thrill of being in this most inhospitable place, they'd soon be tugging at my sleeve and asking to go home. One unusually hot evening in May we decided to make a family trip up there. Determined we would challenge the long-held opinion that the tarn was bottomless, we'd go where no man had gone before: we were going to swim in it. We loaded the quad bike trailer with swimming costumes, towels, snorkels and – as a safety measure – a very long rope.

We drove the bike, now fully laden with children, carefully negotiating the steep slopes and bogs, while the kids chattered nervously about what might be lurking under the tarn's murky surface. I was beginning to share their apprehension, not helped by Clive, who said, 'It'll be a hell of a depth. Yer won't be able to touch t'bottom, an' if yer can then t'mud'll suck thi down.'

I'm used to wild swimming, often taking a dip in the cold waters of the plunge pool beneath the waterfall in Hoods

Bottom Beck; I'd even once swum in Birkdale Tarn. The children, too, love outdoor swimming, and during the summer months, greasy from sheep clipping, we like to cool off in the river – me with a bar of soap to lather on to them, killing two birds with one stone.

'. . . an' then there's the monster, one of them bottom-feeder pike-like things,' Clive carried on, for the benefit of his wide-eyed and terrified offspring.

By the time we arrived at the tarn their enthusiasm for swimming had waned, and they stubbornly refused to put on their swimming costumes. Keeping it up, Clive asserted that he wouldn't go in that water at any price. They all agreed that anyone who ventured in was unlikely to emerge alive.

'It looks like you're gonna have to go in,' Clive said, turning to me. The children seemed to agree that I was disposable, and chucked me my wetsuit. Clive started to make a safety line, tying the rope around my middle, the children all promising they would help pull me out when I got into trouble, as they were sure I would. Reuben cheekily added that we might need to tie the other end of the rope to the quad bike. Reluctantly, I walked to the water's edge.

'Who's gonna look after us?' one of the children murmured. There seemed to be a fair amount of concern about who would make breakfast the next morning.

Tentatively I dipped my toe in the water. It wasn't as cold as I'd thought it would be. I cautiously took another couple of steps. The bottom was soft, but firm. I felt my way along the edge, my feet sinking only slightly into the peaty bed. Clive was getting impatient.

'Ga on,' he shouted. ''Urry up.'

I scowled at him, noting that he'd already lost interest in holding the other end of the rope. I stepped further out, but the water only reached the middle of my shins. I wondered whether I was going to stumble into a deep trench, but on I went, further and further into the tarn. It didn't get any deeper. The whole mysterious, terrifying thing was only fourteen inches deep, no deeper than a paddling pool.

'I allus said that thoo walked on watter,' Clive observed.

The children soon stripped off and were splashing around with me, stirring up all the sediment that had settled on the bottom and rolling happily in the mud. There was no more talk of monsters, but I soon had a whole family of creatures of the black lagoon.

Our water has always come from a spring on the moor, and it is, in its natural state, slightly tinged with peat, but healthy and pure. Mains water has been installed as far as West Stonesdale; beyond that, all the farms and houses are on spring water. Ours used to go straight into a collection tank, then down a pipe to another tank, and then into our house. Occasionally a frog would find its way into the pipe and we'd spend a good half day trying to work out where the blockage was. But the system caused us no real problems, and we were adept at dealing with air locks and leaks.

Unfortunately, our water supply didn't satisfy the Public Health Inspectors. It wasn't to their taste, you might say. Luckily the estate which owns our land, and from whom we rent the farm, footed the bill for a proper water treatment system, which has evolved over the years into something highly technical and beyond the grasp of a layman. It started simply enough with the addition of chlorine to the tank, which turned our bathtub blue and gave my hair a green tinge. It made me wonder what it was doing to our insides: we could certainly taste it. It was also a bit hit and miss: sometimes the tank seemed to get a double dose of chlorine, sometimes none at all. It still had bits of gravel in it.

It was soon upgraded, and now we have a new, much better system, which uses UV filters and does not add any chemicals to the water. The water pressure is much better, which we appreciate when it comes to the shower; but the downside, as I've said, is that the whole system relies on electricity.

Even though we have thousands of pounds' worth of water purification equipment, we still have to be inspected, and the standards seem to be even more rigorous than before. I was busy in the farmyard one morning when the public health inspector turned up.

'Mornin'. What can I do for yer?' I said.

'I'm here to test the water,' she said. 'You have a water tap near the woodshed where walkers and campers can fill their water bottles.'

'Aye.'

'Well, because you are supplying the general public with water, it is a requirement that I test the water quality.'

This annoyed me, because the only reason we installed the tap was to save ourselves the trouble of filling walkers' flasks and bottles in the kitchen. I'd also seen them filling their containers with river water, which I figured was far more likely to cause stomach upsets.

I politely showed her into the kitchen, where she filled her bottles with water samples; then more were filled at the outside tap. Then she set off up the moor to the water tanks and the spring. I decided I was better off staying out of the way, so set about my yard jobs with renewed vigour, muttering occasional expletives about officialdom.

A week later, we got a letter. I'm amazed I actually read it, as I'd assumed it to be a foregone conclusion that it would pass. The letter was full of jargon and figures, but eventually I waded through to the analysis result: 'Your private water supply has failed the quality test.'

We couldn't believe it. I read it again, more closely. It stated that the water that came from the taps was 100 per cent safe and clean, but that the water in the tank at the top of the moor – the one that caught the water as it emerged from the spring in the ground – was dirty.

'Well, who'd 'ave thought it?' I said sarcastically. 'Fancy the water being dirty at source, but clean after it's gone through our vastly expensive treatment plant.'

'Bloody typical,' said Clive. 'What the 'ell is the point of spending a fortune on cleaning t'water if you're gonna fail on it being dirty in t'first place?'

Surely, the only water that matters is the water that comes out of our taps?

'Ah'll tell thi what we're gonna do,' he said, reaching for a red biro. 'Absolutely nowt.'

He circled the word 'private' in the heading, 'Regarding your private water supply'. Then he put the letter back in the prepaid envelope and gave it to the postman the next day.

We had a call from a neighbouring farmer one day to say that two walkers had been going past the Red Gulch Ghyll, and had seen a lamb trapped down there. Raven and I set out to have a look, walking along the other side of the ghyll and occasionally peering down into the bottom. Eventually we spotted a gimmer lamb, half hidden beneath a rocky overhang. She'd probably just reached out in search of some sweeter grass and had taken a tumble over the edge, landing some thirty feet below the pasture on a small ledge, with the beck running through the bottom of the ravine fifty feet below her. All around her the ground was paddled black, so she'd clearly been there for at least a couple of days.

'I'll go and see just how steep the other side is,' said Raven. She went down as sure-footedly as a mountain goat, anchoring herself by grabbing the few clumps of vegetation that had taken root on the rocks. I lost sight of her, and shouted down:

'Where y'at?'

She'd been distracted.

'Mam, yer knaw tha' blue plastic rockin' 'orse we all used to play on? I've found it. It's down 'ere. It must've floated away when t'beck was up.'

'Worrabout t'lamb?'

'Oh, aye, it's still stuck,' she said.

She couldn't get to it, so we headed home and went back the next morning with Bill. This time we intended to get into the ghyll by walking along the beck that ran through the bottom of the valley. We figured that Bill barking from below might force the lamb to look even harder for a foothold to get back up and into the pasture above.

Our rescue mission was in vain, because she was nowhere to be seen. We were sure she hadn't fallen, because we'd have found the body. There was not enough water in the beck to have washed it away. We had to assume that she had scrambled back to safety of her own accord, and been reunited with her mother. We're no longer surprised by the scrapes and ridiculous situations that our sheep can get themselves into, and very occasionally their resourcefulness in extricating themselves, too.

Lambs aren't the only livestock to occasionally go missing at Ravenseat. Our local builders, Steve and Dave, have worked on our barns and outbuildings for years, and they are often accompanied by a couple of terriers belonging to Steve and a dim-witted spaniel belonging to Dave. They were recently up here re-roofing a cow'as (cowhouse) in the middle of what we call Beck Stack, a steep pasture not too far from the farmyard, which we had cleared of sheep. Every day Steve drove up to the barn in his Land Rover, and after the ritual mug of tea, quick study of the *Sun* and much fiddling to get a signal on the battered long-wave radio, work started. Tess, his smooth-coated Jack Russell, and her six-month-old pup Gem, sniffed around, then lazed in the sunshine.

One morning Steve and Dave arrived with a cargo of scaffolding poles to shore up the west wall of the building. Steve drove his Land Rover up the field and let the two dogs out. Then he dropped the side of the trailer and the metal poles rolled out onto the ground with a deafening crash.

Tess took fright and bolted, her short tail tucked between her legs. Gem scarpered too, but nobody felt alarmed. Steve and Dave were sure the dogs would be back by the time they finished work. But when they drove back into our yard, Steve announced, 'We've a dog wantin'.'

Tess was back in the Land Rover, curled up on the front seat, but Gem had not been seen since the scaffolding pole incident.

'Right, children,' I shouted, summoning the troops. 'We've a dog to find. A small white 'un, last seen in t'Beck Stack near t'cow'as.'

We formed separate search parties, confident she couldn't have gone far. There's nothing that the children like more than a mission: the challenge was simple, find Gem. Steve walked along the beck edge, whistling for her. Dave headed upstream, looking across into the seaves along the bankside. The children went in all directions, walking, running, Reuben cycling up the shooting track. No luck.

The next plan was for Steve to drive back home to Reeth, eighteen miles away, calling at our neighbouring farms in case she'd wandered further than we thought, and also dropping in at the campsite in Keld in case she'd been picked up by passing walkers.

At darkening, Steve drove back into the yard.

'Nae, I din't see t'point in sittin' in t'chair wonderin' abaht it,' he said. 'Thought I'd 'ave a ride back up t'dale an' 'ave another look.'

'Could she 'ave gone down a rabbit hole?' I said.

'Mebbe gone down t'beck,' said Clive. I scowled at him.

'No idea where she's at,' said Steve. 'Last time I see'd 'er she was wi' Tess, just afore t'poles come off t'trailer.'

He went home, empty-handed and heavy-hearted.

Clive and I were in the yard early the next day, feeding the tup hoggs, when Steve rolled up. He wound his window down.

'Yer early,' I said. 'Dave's not 'ere yet.'

'I thought that me an' Tess would 'ave another look around for Gem,' he said.

'Come here an' 'ave a look at this tup hogg,' said Clive, catching one of the sheep by the horns. 'I'm wonderin' whether 'e's a good 'un.'

Steve switched off the engine and walked across to where Clive was wrestling with the hogg. 'Let's 'ave a look at what 'e turns up like,' said Steve.

Clive rolled the tup over and sat him up, holding him steady so that we could see his tackle.

'Grand, in't 'e,' said Clive.

There was a moment of silence. Steve was deep in thought, and we waited with bated breath to hear his verdict on the tup. He frowned, then turned to me and gestured for me to be quiet. In the silence I, too, heard a faint whimpering noise.

'What were that?' said Steve, turning in the direction the sound came from, towards the Land Rover.

'Yer what?' said Clive, who was straining to hold the wriggling tup.

The noise seemed to be coming from the vehicle. Peering through the windows, all we saw was Tess dozing on the front seat. Steve opened the door and fumbled under the dashboard to release the bonnet catch. As he lifted the bonnet, the whining noise became louder.

'Bloody 'ell! Come and 'ave a look at this,' Steve exclaimed.

Peering upwards, eyes wide, was Gem, jammed in the chassis framework next to the engine block.

'I can't believe that! T'lal' bugger 'as been in there all this time.'

We watched as he extricated her. She was dirty and shivering with fear, but otherwise unscathed. It was nothing short of a miracle that she had escaped injury. She must have run under the Landie when she was frightened, and she'd been there ever since, making four trips totalling nearly eighty miles – not counting the messing about going to the other farms and the campsite. Plus she'd been there overnight. It's amazing that she didn't fall out, or get burned on the engine. She certainly was an extremely lucky dog.

By the end of May, all our yows and single lambs have been returned to the moor. The yows with twins are still in the pastures and allotments. Sheep are now banned from the meadows, to give the grass a chance to grow, and Clive is on the lookout: any sheep that dares to sneak over the cattle grid or make its way under a water-rail will be skedaddled back pronto, and persistent offenders have been known to be delivered to Hawes auction mart the following Tuesday.

'I'll warm 'er arse, she's a bloody flying machine,' he says as he and Bill set off to remove another trespasser.

We don't use any chemicals or fertilizers on our meadows, and we let them reseed themselves naturally. The only thing we give them is a dose of muck. Hay is a vital crop for us: how much we manage to grow makes a huge difference to our feed bills in the winter.

Walking the meadow walls is a pleasant but time-consuming job. We pick up any topstones that have been dislodged, keeping the boundary walls in order and also preventing mishaps with the mowing machines when we harvest the grass. An encounter with a big, heavy rock brings mowing to an abrupt halt.

Occasionally at this time of year we are inspired to have a go at growing other things, apart from grass. We've tried potatoes, green beans, carrots and salad leaves. The children enjoy planting them and cultivating the plot, but the harvest is never worth the effort: we get a few misshapen veg if we are lucky. Reuben has a small vegetable and herb patch down by the graveyard, which he's fenced off to keep out intruders like horses, rabbits, sheep and chickens. We suspect Reuben was motivated more by the chance of unearthing buried treasure or a Viking hoard than becoming the next Percy Thrower.

During the war, when everyone was told to 'dig for victory', there were efforts to grow vegetables up Swaledale. But even in those desperate times they weren't very successful, and all our attempts have more or less proved that this just isn't the right terrain. We're high, windy, bleak, cold; the soil is peaty, acidic and often sodden. We did manage to grow cucumbers, although maybe I shouldn't use the plural, as we actually only got one and it was the furriest cucumber I have ever seen. It had clearly adapted to life at Ravenseat, and it made me laugh just looking at it. Not much can tolerate these conditions: grass, pignuts, microscopic bilberries – and rhubarb. Rhubarb thrives. We have a very productive rhubarb patch, and luckily rhubarb crumbles are very popular with Clive and the children.

Our friend Alec, who had joined us for tea, expressed his long-held theory that rhubarb causes impotence.

'Causes what?' said Clive. 'Din't catch that. What's 'e sayin', Mand?'

I didn't bother repeating it, as I dished up rhubarb crumble to the seven children sitting round the table.

We don't live anywhere near the famous Rhubarb Triangle (around Wakefield in West Yorkshire), but it still grows in such abundance here that for a while I sold crates of it to a high-quality jam and preserve company. They were very impressed by the quality and quantity that we produced and asked if we used any special fertilizers. I didn't mention that the area where it was harvested was above the septic tank, and in a field called the graveyard (an area behind the woodshed, which used to be an old chapel).

Trade for store and breeding cattle can be good in May, with special sales held at the auction marts. It is a time when we normally sell some of our stirks, our young cattle; but recently Clive has been buying more young stock to graze in our allotments during the summer. He set off one morning with an empty trailer and the chequebook, to go to the auction in Carlisle to buy what was in his price range. I rang for an update, and he told me it was a buyer's market, the cattle were good, the prices low, and he'd just kept on buying them . . .

''Ow many has ta getten?'

'Thirteen,' he said, sounding pleased with himself.

'Crikey, thoo'll nivver get 'em all in t'trailer.'

'They were cheap, I just kept flappin' mi 'and,' he said.

'Well, that's grand. But I's thinkin' that yer should flap thi 'and yance more, thirteen being unlucky an' all that,' I said.

'I nivver 'ad yer down as suspicious,' he said.

'Superstitious,' I corrected.

'Whatever. Anyways, I'm nit buying any more coos as I'm off to t'canteen. There's a nice Polish lady there who allus gives me an extra dollop o' mash on mi plate.'

The phone rang when I was clearing up after teatime, with Clive's meal in the bottom oven waiting for his return.

'Hello, can I speak to Mr Clive Owen?' said a man with a thick Scottish accent.

'No, I'm sorry but he's at the auction today. Can I help or give him a message?'

'Yes, you can tell the wee laddie that he's mebbe a coo short.'

The man, a drover, was looking at a cow standing forlorn and alone in the lairage at the auction mart. The ear tags corresponded with paperwork in the office that said Clive had bought her earlier in the day.

I rang Clive on his mobile and asked him the obvious questions. How many cows had he bought, and how many did he have in his trailer?

'I bought thirteen, an' I've got 'em all on board,' he said.

'Where are yer at?'

'Kirkby Stephen. I'll be 'ome soon,' he said.

Half an hour later, he pulled into the farmyard.

'C'mon, let's get 'em unloaded, they'll be tight. We'll run 'em into t'pens,' he said as he pulled the trailer's rear door down.

It was by now dusk but even in the fading light there was no mistaking the fact that only twelve cows came out of that trailer. Clive was dumbfounded. The drovers had run the cattle down the alley and onto the loading docks for him, but somehow one must have found its way into another person's pen and been left behind.

'It looks like yer off back to Carlisle.' I said. 'Yer can tek yer tea with yer.'

Doing the one-hundred-mile round trip twice in one day left a sour taste in Clive's mouth, but it wasn't enough to put him off Carlisle auction mart for good. A reduction sale of a re-nowned herd of pedigree Beef Shorthorns was too tempting to miss. He sent for the catalogue, and off he went again. This time I knew he wouldn't be filling his trailer, as these cows would be expensive. Beef Shorthorns have become a fashionable breed and there would be plenty of buyers for them. By the time Clive and his farming friend Mark got there the sale had begun, and the ringside was packed three deep with buyers. Folks had travelled from as far away as Ireland just to be in with a chance of buying in some of this famous bloodline.

Clive had set his heart on buying an in-calf heifer: two for the price of one, he reckoned. He was determined not to come home empty-handed. Ballyliney Lancaster Zabrina was the heifer who took his fancy, and he waited for her to come into the sale ring. The auctioneer set her off at a thousand pounds but she didn't stay there for long; the bids kept coming, the price getting higher in ever increasing increments. After what seemed like an age, the hammer went down at £3,000, and she was ours. I rang Clive for an update.

'I've getten a coo,' he reported. ''An I's off to t'canteen now, 'cos . . .'

'I know, I know,' I said. 'Just mek sure that yer remember to put yer coo in t'trailer.'

I got my first sight of Zabrina when the pickup pulled into the yard, her wild eyes peering back at me through the ventila-tion flaps. Ravenseat was now her home, and she joined our little herd of Beef Shorthorns. After the usual bit of argy-bargy among the cows, she was accepted into the herd. There's always a bit of tussling when a new animal is introduced to a herd or flock. You sometimes get an almighty fight (or sometimes an orgy), but fortunately for Zabrina, in her delicate state, it was all very tame. Order was restored, and they were soon all grazing quietly together. A few weeks went by and her calving date was drawing near. Not wanting to risk her taking off onto the high ground to calve, and as it was her first calf, we decided to put her in a barn so that she could be monitored. Every day we'd both have a study of her.

'Aye, she's near her calf now,' Clive would say.

'Look at t'size of 'er. Any day now,' I'd agree.

This went on for days, Zabrina getting ever more round.

'Crikey, she's gonna have a big calf.' I'd wince and rub my own bump in sympathy.

Finally the day came. Zabrina's tail was held high and she occasionally glanced backwards at her own distended belly; she'd panch – pacing to and fro, never settling. Clive panched too – he couldn't settle any more than she could. It was 4.30 a.m. when he came into the bedroom and woke me up: she was calving and he needed another pair of hands.

Native breeds tend to have easy calvings and shouldn't need too much help, but Clive wasn't taking any chances. This cow was precious, and the signs were that she was carrying a big calf. She hadn't lost her appetite, and it was simple enough to tempt her to the feed trough and slip a halter over her head. You're not safe in a confined space with a sizeable animal like that charging around, and a cow in labour is not going to be the most reasonable creature (any more than a woman is). We wrapped the rope around a solid metal post, and I held the end while

Clive checked to see whether the calf was correctly presented: two front feet and a nose in a diving position.

'It's coming reet, but it's got a fair old pair o' feet on it, she'll struggle to calve it herself,' he said.

Next he attached a long length of the thicker baler twine that you get on the Hesston size straw bales, tying an end to each of the calf's front legs just above the hooves. We have in the past used a calving aid, a ratchet-type contraption that can be used to help a delivery; but although this is a useful tool in the right hands, it can also cause great damage to the cow and calf. So for us it's back to the old favourites: baler twine and muscle power. Gentle and constant pressure on the twine at a downward angle helps to ease things along, but the danger can be that the cow will go down at this point – which is why she must not have her halter too tightly secured or her head trapped in any way, as this could impede her breathing or even strangle her.

Zabrina decided to lie down, and I loosened my hold on the rope, quickly tying it to the post but leaving plenty of slack. Then I went round to the business end, where Clive was now sitting on the straw with the string taut around his back. The hooves and nose were out; Zabrina was straining and emitting some pretty harrowing bellows. Suddenly the calf's head appeared, wet and steaming in the early-morning air. Now it was time to pull with all our might, a matter of urgency to try and keep the calf moving, the shoulders and hind quarters being the most likely sticking places. The calf began to come away, and soon the forequarters were out. Clive was sweating, but all was going well.

'Nearly there,' he panted.

The long length of the calf's body was now laid out behind the cow on the straw.

'Keep pullin', come on, Mand, it's not movin', he's fast at his hooks.'

Sometimes a calf's hips stick within its mother's pelvis, bone to bone, and pulling gets you nowhere. It's a very bad thing to have a calf stuck at this point, and it's surprising how quickly the life drains out of them. We changed positions, me now pulling the calf's legs while Clive attempted to slightly rotate him – not easy, as he was a big calf, heavy and slippery. Clive managed a slight twist, and that was all it took: with one final push, Zabrina's calf was born. He lay outstretched and motionless on the straw. Clive slapped his chest with the flat of his hand. I squeezed his nose and muzzle to remove the mucus and pushed a blade of straw up his nostril to make him sneeze, but there was no reaction.

Clive rummaged in his kitle's top pocket, and among the scraps of paper, pennies and fluffy mints he found a vial of Dop drops (Dopram V). This medicine, a respiratory stimulant, is an absolute lifesaver. A few drops under the tongue of a newborn struggling to live will cause a sharp intake of breath. It only works in those first crucial moments after birth, and thankfully Clive had it on him. Prising the newborn's mouth open he tipped in the whole contents of the bottle: an overdose by any standards, but desperate times call for desperate measures. We both watched, holding our breath, willing the calf to cough. Thankfully, within seconds the Dop drops took effect. He gasped, and then blinked.

Carefully I untied the twine from the calf's legs whilst Clive removed Zabrina's halter. Zabrina seemed to have gone into a trance. Sitting on her haunches, she seemed none the worse for her experience, but she had not investigated what had caused her such pain. She seemed to be in denial, and we decided that the sooner we departed and left her to her new role as a mother, the better. We carefully pulled the calf around to her front so that she could not ignore his presence. By the time we were at the barn door the calf had lifted his head, and Zabrina was looking a little more interested in him. She soon came to her senses and when we went back to check on progress, both were standing. It would have been tremendous to have had a heifer calf to breed from, but we knew we were lucky just to have a live calf.

We occasionally need to call in a vet. Where possible we take the patient to the vet's at Kirkby Stephen, but with the cows and horses, the vet has to come to us. A few years ago now, when we kept more continental cattle, one of our cows gave birth to a bull calf. The cow had calved easily enough, but shortly after the birth we noted that the calf's front hooves were knuckled over. Although it's not a common condition, I had seen it before: we once had a foal born with the same problem, but not quite so severe, and only on one of its hooves. The foal was quite mobile, and eventually over time it corrected itself without any intervention.

But for this calf it was more serious. The cow was very attentive to her newborn, and after licking him vigorously she nudged him to encourage him to stand – but his front legs would not straighten enough to allow him to get to his feet. He tried, but his knees buckled, and he collapsed into a heap with his legs splayed. His desperation to stand, and his mother's helplessness and distress at being unable to feed him, were pitiful to see. All we could do was milk her and then decant the warm colostrum into a feeding bottle. As I sat in the straw alongside the calf, his neck stretched out across my legs and head tipped back sucking the bottle, I wondered what, if anything, could be done.

'I'll ring t'veterinary,' said Clive. 'See what 'e thinks.'

He talked at length with our vet, describing the calf's deformity.

'I'll consult with a colleague as to what's the best course of action,' the vet said. 'I'll ring you back shortly.'

Clive put the kettle on while we waited.

'In all mi days, I've nivver seen one as badly affected as this 'un,' he said as he brewed the tea. 'I just don't know whether them legs will ever straighten.'

'Poor l'al beggar,' I said sadly, finding solace in the biscuit tin. 'I 'ate it when summat like this 'appens.'

We hadn't finished our tea when the phone rang. The vet, having consulted a colleague, felt that surgery was the correct course of action, and we were referred to a specialist equine veterinary practice that had experience of dealing with severe neonatal bilateral flexor tendon contracture in foals.

'It's a simple enough procedure,' our vet assured us.

Later that day we took the calf to the surgery, where we were expected. We were to wait while the operation took place, and then bring our calf back home with us once the anaesthetic wore off.

The cow was naturally not keen to part with her baby, venting her displeasure by bellowing loudly as we whisked the calf away and laid him out in the back of the pickup.

'We'll 'ave 'im back to yer in a crack,' I shouted to her, and I had no reason to think otherwise.

It was a fair trek out for us, and we were relieved when we finally saw the sign for the equine clinic. Relief soon turned to unease, though, as we drove past signs for the hydrotherapy pool and the solarium, and past the rotary walker and exercise area. This place was seriously fancy, more like a spa hotel than the bog-standard veterinary surgery we were used to. Eventually we found our way to reception, where we were greeted by a jodhpur-clad receptionist.

'Ahh, now you must be Mr and Mrs Owen. You've brought us

a calf, haven't you?' she said. 'If you'd like to take a seat, Bruce will be with you in a moment.'

We sat down, Clive peering into a highly illuminated fish tank whilst I flicked through the magazines: *Equestrian Life*, *Horse & Hound* and suchlike.

'It's all a bit posh 'ere, in't it?' I whispered to Clive.

'I bet this is gonna be expensive,' muttered Clive – rather too loudly, I thought, as the receptionist eyed us up from the other side of her desk, her gaze coming to rest on our wellies, complete with clarts of mud stuck to their thick soles.

We didn't have long to wait before Bruce put in an appearance, and he was everything that I would have imagined a horse vet to be. Cutting rather a dash in a tweed blazer, moleskin trousers and dealer boots, he was a far cry from the vets with rubber-coated aprons and waterproof leggings who we usually dealt with. After the introductions, it was down to business.

'And where is the patient?' he asked.

We traipsed across the car park and showed him the calf lying on a bed of straw in the pickup.

'This is one of the most severe cases I've ever seen,' he said, scratching his head. 'But I think I can sort it out.'

We carried the calf to the back of the clinic and into the large animal operating theatre, lying him on the heavily padded floor whilst Bruce scrubbed up. No scrubbing brush and bucket of cold water here: it was all very clinical. I stroked the calf's head, holding him still while Bruce snipped away a small patch of hair on his neck where he could raise a vein. Then he filled a syringe from a bottle, glancing at the calf; holding the syringe aloft, he flicked it with his finger to disperse any air bubbles.

'I'm going to have to ask you to stand aside now,' he said to me as he administered the anaesthetic. 'I like things to be kept sterile.'

He looked towards me and raised his eyebrows. I stepped back to stand next to Clive, who was shuffling uncomfortably. The calf was soon asleep.

'We're going to be watching a master at work,' I whispered to Clive as Bruce assembled all his equipment. There was a stainless-steel trolley with all manner of medical accoutrements arrayed neatly on top.

After covering the calf's prostrate body with a green sheet and wiping its front legs with an iodine-based antiseptic, he switched on his head torch and reached for a scalpel. He talked us through the whole procedure, telling us how he was going to make a succession of small incisions along the length of the tendons, hopefully allowing them to stretch enough to be able to straighten the joint. Then he would splint the legs, bandaging them tightly in a straight position, and for a while the calf would need assistance to stand. Within a few weeks the tendons would heal, the bandages would come off and we would have an upright and mobile calf.

It sounded relatively simple and Bruce set to work, on his knees, hunched over the calf. There was no blood, not much to see at all. Every so often there'd be a metallic clank as the vet put the scalpel on the metal trolley and reached for the surgical scissors. There was a bit of manoeuvring of the patient when work commenced on the second leg, but eventually the suturing began.

'I need your assistance now,' Bruce said, looking towards me. 'I'd like you to hold his leg straight while I bandage.'

He padded out the plastic splint with cotton wool, and then rolled out reams of vetwrap. Switching off his head torch, he got to his feet and momentarily stood admiring his handiwork.

'That's it,' he said, reaching for the green sheet that covered the calf. 'Now all we have to do is wake him up.'

As Bruce lifted the sheet the expression on his face changed in an instant. He dropped to his knees, hurriedly putting his stethoscope on the calf's chest, moved it a little, then looked up to Clive and me and said, 'Erm . . . I'm afraid he's dead.'

Clive and I stood open-mouthed as Bruce took off his head torch and went to wash his hands. Lying in front of us was a perfectly healthy-looking calf, with straight legs. But dead.

'There'll be no charge for this,' shouted Bruce, drying his hands. 'I'll give you a lift with him back into your motor.'

We were so shocked that we never got to the bottom of what had happened. It was only really Clive's words of wisdom on the way home that helped numb the pain.

'Mand, 'e just wasn't a reet 'un. Yer know, there could 'ave been far more wrong with 'im than met the eye. We'll get 'is mother another 'un, foster one on.'

Fortunately, bad experiences are few and far between. Sad times are part of the fabric of our lives, and in any day our emotions can veer from pure elation to despair, but it's important to make the best of a situation and learn from it. We gave the cow another calf and, darling that she was, she accepted it without too much trouble. But that was the last time we ventured away from our usual veterinary surgery.

Clive once bought a batch of six calves quite cheaply, owing to one of them having a noticeably large umbilical hernia. They lived together in the barn on a diet of powdered formula milk and dried feed. Time went on, and they all thrived: long-legged and shiny-coated, they'd buck and race around the pen. Only the one with the hernia was quieter, standing to one side while the fun and games took place. We spent many evenings just leaning over the cattle feeder barrier, watching them larking about. Not only was it a pleasure to see them, it was also a good time to make decisions: should we up their feed, should we start weaning them from milk and introduce some hay? That is what stockmanship involves – observing your animals. It takes time and experience to become a good stockman, and it starts with looking.

'I wonder if that 'ernia is botherin' 'im?' Clive said.

'What – strangulating summat?' I said.

Climbing over the barrier, we cornered the calf, who, having seen us coming, forgot his listlessness and became a whole lot more agile. We both felt the tennis-ball-sized lump underneath. There didn't seem to be anything really worrying; it wasn't festering. But unsurprisingly, the calf didn't seem happy about us poking it.

'It's definitely uncomfortable,' I said, stating the obvious.

'I think that we should talk to t'vet,' said Clive. 'I'll mention it when I'm next at Kirkby.'

The following Wednesday we loaded the calf into the little trailer and headed for the surgery, our vet having told Clive he had devised his very own non-surgical (almost) method for dealing with umbilical hernias. I must admit to being slightly on edge after our previous calf operation, but it was obvious that something needed to be done. The hernia was not going to go away by itself.

When we got to the vet's I sent Edith into the small trailer to get him out – easier than getting down on my hands and knees and clambering in there myself. She pushed from behind and our calf emerged, looking startled. Before he had time to escape across the car park I deftly slipped a halter over his head, and with me pulling and Edith pushing the reluctant patient arrived in the garage that adjoins the surgery, which was to be our operating theatre. Within seconds of us arriving the vet had administered the anaesthetic. The calf rocked a little, then sank to his knees, and with one almighty push we rolled him onto his side. Our trusty vet produced a long steel rod and a large set of wire cutters: so much for state-of-the-art surgery. It looked as if he was going to be doing a bit of DIY.

Edith sat down on an upturned bucket, her elbows on her knees, watching closely with her face resting in her hands. Sidney and Violet crouched down to get a better view, hoping there would be a gory story to recount to their brothers and sisters. I held the halter, intrigued to see what was going to be done, half expecting that there'd be viscera or giblets spilling out when he made an incision into the taut skin that covered the bulging hernia. The vet had by now cut the steel rod into two short pieces of equal length, and kneeling down, he took hold of the hernia and pushed it until it went back inside the calf. Nipping the loose skin together, he made four incisions, pushing one of the steel rods through two incisions at opposite corners and leaving a couple of inches protruding at each end. Then he did the same the other way, thus making a cross shape with the steel rods. He took a roll of cord from the floor beside him, and wound it round and round the crossed rods in a weaving pattern.

'I din't know that thoo were into macramé,' I said, fascinated by what he was doing.

'This,' he said proudly, 'is my own patented – well, actually, it's not patented – method for repairing umbilical hernias. It's simple really. The steel pins provide an anchor to cover the gap in the muscle wall; wrapping the cord around secures it and covers the area that the pins do not. The strangulated skin to which the blood supply has been cut off sloughs off and the hernia remains inside, due to the build-up of scar tissue that covers the hole.'

'Is that it?' I asked.

'Aye, that's it,' he said, tying off the cord, anchoring it securely to the rods. 'Give him a shot of penicillin to ward off infection, keep the wound clean and make sure his pen's nicely bedded up.'

And with that he was gone, to attend to his next four-legged patient. By now the calf was stirring. We'd been told to keep him lying down for a short time to let the anaesthetic wear off, and to give him time to come to his senses before trying to get him to his feet. Edith stroked the calf's neck gently and I held his head firmly to the floor, while Sidney and Violet scrutinized the vet's handiwork, disappointed that the whole procedure had been quite tame and involved more in the way of knitting than disembowelment. We were soon homeward bound with our calf, and were delighted when a few days later he seemed to have brightened up and was getting more involved in the rumbustious activities of his counterparts. The rods, cord and dead skin dropped off as expected, and were last seen being carried off triumphantly in the jaws of Chalky the terrier.

The terriers, Chalky and Pippen, are usually to be found patrolling the yard, lazing in the sunshine or perhaps loitering around the picnic benches looking up at the walkers with sorrowful eyes, willing them to part with crisps or scones or indeed anything edible. Everything is on the menu where Chalky is concerned, and she's been known to stick her head into a temporarily discarded rucksack and help herself to a sandwich, wrapping and all.

On first acquaintance Pippen could be seen as the underdog, quieter and shyer than the bouncy, energetic and ebullient Chalky. But Pippen is just older and wiser, and prefers to remain in the background. She's no soft touch, for under that brown and black wiry-coated exterior she harbours a deep hatred for rats, mice and the gamekeeper who lives in the cottage next door. The first two I can well understand, but the gamekeeper issue has had me perplexed for a long time. The minute she hears his Land Rover engine her ears prick up, her hackles rise and she's after him at full speed, biting at his wheels until he accelerates away and leaves her angry and snarling in the middle of the road. It's nothing personal with Pippen, it's all-encompassing. She hates all gamekeepers.

One that we knew quite well would occasionally call upon us.

'Wolfy's here,' the children would chorus when he appeared in the yard.

'I know,' I would shout, because Pippen would be emitting a deep, rumbling growl behind the door, ready to pounce on him the moment it was opened.

Wolfy was a gamekeeper on the estate, and also a keen amateur photographer whose ambition was to have one of his photographs published on Page Three. Not Page Three of the *Sun*, though; this was Page Three of some monthly gamekeepers' magazine. The idea was that you would take a picture of the largest, most impressive dead animal, pictured with the sweetest child that you could find. Wolfy could supply the first, and he looked to me to supply the small, innocent-looking child to pose beside the corpse.

'Come in, Wolfy,' I'd shout. The door would fly open and he'd come loping in.

'I wonder if I could borrow Raven,' he'd say, ignoring Pippen, who by now was hanging on to his gaiter. 'I've gotten a helluva mink,' he'd add. 'A beast of a thing.'

When Raven got too big for the job, Reuben took her place – the whole idea being that the smaller the child, the bigger the verminous animal would look. I did worry about the glorification of the animals' demise but I am not fond of crows, rats and their ilk, having seen at first hand the damage they cause. I did question whether it was entirely normal to want to take pictures of dead animals, though, especially when Wolfy brought his holiday snaps with him to show us.

'This is me skiing,' he'd say. 'This is a picture of the resort where I stopped.' So far, so normal. 'An' this is a dead beaver that I found at side o' t'road.'

'Nice one for t'album,' I'd say.

At the end of May, on the last Thursday of the month, there is one of the first of the Swaledale sheep shows: Tan Hill Show. Tan Hill Inn is our 'local' pub and is famous for being the highest in England, at 1,732 feet above sea level. The pub has been there since the seventeenth century, although there are records of a public house being on the site for at least a century before, set up to quench the thirst of the men digging for coal at Tan Hill Pit. Tan Hill coal was of notoriously poor quality, but locals used it, as it did burn hotter than the peat fires to which they were accustomed.

A couple of evenings before show day, we and other local farmers in the area gather at the pub to construct the sheep pens, using wooden hurdles to make the small pens on the open moor around the pub. Our friend and neighbour Clifford Harker was one of the founder members, and a great stalwart of the show. He was at the very first show meeting in 1951, was show secretary for twenty-five years, president for four years and attended every show and every meeting until his death. Every year, as an act of remembrance, Clive takes Clifford's rubber mell with him to hammer in the posts.

Hill shepherds have just come through the toughest, loneliest time of the year, tending to their lambing flocks, and Tan Hill Show is greatly anticipated as it brings with it an opportunity to catch up on the local gossip, meet old friends and, of course, indulge in the odd pint of beer. The show itself is very much aimed at the Swaledale sheep purists, with little else to distract from the important business of judging the sheep.

The Tan Hill Inn was famously used in an advertisement for a well-known double-glazing company, the incessant wind and rain that batter this desolate and exposed place making it perfect to showcase their draught-free windows. Although it is held in late May, show day is almost always blustery and wet. A catering van supplies hot drinks to warm gnarled hands that are now numbed by the cold, the sheep breeders unwilling to leave the showfield for the pub for fear of missing a judge's critical decision. A stall sells coats, boots and overtrousers to those who, after a whole winter dressed from head to foot in waterproofs, perhaps shed too many layers when tidying themselves up for the day.

They say that at one Tan Hill Show in the early eighties, thirteen gallons of whisky were drunk. The festivities go on late into the night, with plenty of deliberating as to whether the judges were right and speculation as to whose sheep might be making the money at the back end sales. And you can be sure that the following day, there'll be a few sore heads in the district.

6

June

Sometimes I think back to the days before I worked full-time on a farm with a clutch of children to care for. Although I wouldn't swap my life for anyone's, I have flashbacks to being a teenager dancing all night in the clubs of Huddersfield: Calisto's, the Plaza, the K.U. I'd dance till dawn and then make my way home slightly the worse for wear, the music still ringing in my ears. I was never a hardened clubber, but I enjoyed the music and I've always loved to dance. I'd leave home dressed reasonably sensibly, then call at a friend's house and change into something skimpier that fitted where it touched. It was all a far cry from my life now, but they are good memories to have. And that's what they should have stayed: memories . . .

Clive's friend Steven is a farmer and sheep breeder, but in the 1990s he ran a club night called Hard Times, which started in Leeds but played other venues around Yorkshire, including Huddersfield. He was the promoter, and he lined up some top-notch DJs who manned the mixing decks playing dance music while hordes of teenagers and twenty-somethings danced all night. I remembered when Hard Times was a big fixture in the calendar of Huddersfield youth, and how I would pick up its distinctive flyers in Fourth Wave, the grungy music shop on the fringe of the town centre that catered for would-be hipsters.

It was the Queen's Diamond Jubilee weekend in June 2012, and Steven decided to hold a Hard Times Nineties revival rave at the Warehouse club in Leeds on the Saturday night. We talked about it the week before, when we were in the sheep pens sorting up the tup hoggs.

'Aww, I used to love clubbing,' I said casually, as Steven, Clive and I leaned on the gate. 'Happy times, dancing til't next morning . . .' I added wistfully.

'I used to go dancin' on a Saturday night,' said Clive. 'There was always music at t'Black Swan at Ravenstonedale. Nancy sometimes used to bring 'er squeezebox.'

'Not the same thing,' I said.

'Definitely not the same thing,' said Steven, shaking his head. 'Nah then, I tell yer summat, I'm gonna give you a ticket, not just any ticket, a VIP golden ticket, an access all areas ticket wi' complimentary drinks thrown in.'

'That'd be just proper,' I said, smiling. 'I can't wait.'

Clive just raised his eyebrows and half smiled to himself. 'I bet yer won't gan,' he said.

That was it: the gauntlet was thrown down.

'I will go, it'll be just like the old days,' I said, trying to convince myself too. 'Count me in, Steven, I'll be there.'

For old times' sake I was going to sample the bright lights of the big city once again. I'd always had a brilliant time clubbing, and in my head it was going to be a re-run of those happy, carefree nights.

For the next couple of days Clive would say, plaintively, 'You're not really going, are you?'

'Do you mind?'

'Nah, not at all . . . I just think that yer as green as the hills
that thi comes off, yer might get mugged or murdered.'

'Yer forgettin' summat, Clive,' I retorted. 'I don't come off no
hill.'

I was hardly deserting him: I would do my jobs, serve the
afternoon cream teas and then get on the train at Garsdale, the
quietest railway station you could imagine, overlooked by a
small field of donkeys. Then a couple of hours later I'd be in
Leeds city centre, where I would check into the cheapest hotel
I could find and change into my party clothes. Then I'd walk to
the club, dance until the wee hours, come back to the hotel
to get changed back into respectable attire, and catch the train
back home. I'd be back at Ravenseat to do my jobs and bake by
10 a.m.

And, remarkably, that is exactly what happened: it all went
without a hitch. I was between pregnancies, so nice and slim.
Once in my hotel room I dumped the farm clothes and squeezed
myself into some exceptionally tight skinny jeans, a leather lace-
up bustier (I struggled with this, as there was no one to fasten it
up) and vertiginous heels. After a liberal application of make-up
and a lot of backcombing, I headed for the club.

The queue of people snaked along the side street. I decided
against pushing to the front of the queue wielding my VIP
ticket, as I had a feeling I could end up getting thumped for
queue-jumping. It felt like stepping back twenty years, shivering in the night air, enjoying the sights and sounds of a city. The
thump of the muffled bass from inside was punctuated by the
wail of police sirens, shouts and constant traffic noise. It felt so
familiar, yet my senses seemed heightened by my years away.
Once I'd finally got into the club I pushed through the throng
and made my way towards the upper floor and the members'
area, where I could sit and survey the scene. Sipping a glass of

champagne, part of my VIP package, I looked down on a quiet dance floor. There were a few groups of people milling around the stage decks, but plenty of folk crowded along the blue-lit length of the bar. I must have been sitting for some time before I got talking to a group of girls who were on a night out. When I say talking, I really mean shouting and gesticulating above the music. I had a sinking feeling that I was becoming my own mother, as phrases like 'I can't 'ear me'self think' popped into my head.

'Where are yer frae?' my new-found friends shouted.

'Ravenseat.'

They looked blank.

'Swaledale?' I tried. They shrugged. I wasn't sure whether they didn't know it, or whether they just couldn't hear.

''Uddersfield originally,' I said. They nodded.

By now the dance floor was filling, and I decided I could get down there and in amongst it without embarrassing myself. Hours later I was still there, hemmed in by a sea of people, arms waving, whoops of joy going up when an anthem came on. The strobe lights cut through the smoke that hung in the air; I could smell the drink, the sweat, and feel the heat of the press of bodies. It seemed that all the people and the life I had left behind me were there, never having moved on. I saw faces I thought I recognized, their features momentarily illuminated by the blinking lights. Maybe some were on nostalgia trips, taking a step back in time like me, but it felt as if I was in a time warp. Right then I knew my days of clubbing were over. I was having fun, but revisiting the past reinforced in my mind the fact that my heart was at Ravenseat, far away from the city noise, bustle and chaos.

I danced until 4 a.m. and then walked barefooted back to the hotel, ceremoniously dumping the heels in a recycling shoe bank that I spotted in a supermarket car park. High heels were never a good look for me: standing at six foot two in my bare feet means that when I wear them I tower above everybody, perhaps raising suspicions that I am a bloke, a cross-dresser. My feet, used to being in wellies, were also killing me. I remember thinking that living in a boggy place means it will always be the people with the widest feet who won't sink in and will therefore survive: I'm perfectly evolved for Ravenseat, I guess.

I rang the bell at the hotel entrance, and the night watchman let me back in. His face was blank and expressionless and we never made eye contact: shoeless woman, looking worse for wear, rolling in at 4.30 a.m. – he'd seen it all before. Back in my room, my ears still buzzing with music, it was time to change back into shepherdess mode and get myself onto the 6.30 a.m. early train back to the sticks. The train was far busier than I expected: the majority of the passengers were clearly revellers who were now feeling the after-effects of their night. Some slept, the train guard having difficulty rousing them when they reached their stops; others talked, shouting as if their ears hadn't yet adjusted to normal. I'd always wondered why the trains on the Settle to Carlisle line had a dubious reputation, and it became clear as I watched a retching youth hurry up the centre aisle towards the toilet. I just stared out of the window, watched the sunrise and the beginning of what promised to be a glorious day. As the train neared home the scenery became familiar: the drystone walls, the sheep grazing quietly in the fields, a perfect pastoral vision in a world that I was glad to call my own.

Back on the farm the bigger children were impressed, wanting to know what my night had been like. Clive was impressed too.

'Yer some woman, thee,' he said. 'I din't think that yer'd 'ave t'bottle to do it.'

I'd gone up in Clive's estimation, and, even better, he'd also recognized how much work goes into getting the children dressed and breakfasted of a morning.

'I'm soooo tired, but I'd better get baking,' I said, playing the martyr card.

'Aye, sun's gonna shine an' I dare say that we're gonna 'ave a few visitors today,' he said.

Having missed a whole night's sleep, my eyes were red-rimmed and I was blinking from a combination of tiredness and the smokiness of the venue.

''Ow do I look?' I asked Clive.

'Rough,' he said, not mincing his words. 'Yer eyes look like piss 'oles in t'snow.'

The bell rang, and Reuben answered the door.

'Two cream teas, Mam,' he hollered.

Putting up the order, I set off on what felt like it was going to be a very long day. Down to the picnic benches I went, treading carefully to avoid the terriers and squinting in the bright sunshine. Reuben was, as usual, entertaining the customers.

'Here we are, two cream teas. All freshly baked this morning,' I said as I slid the laden tray onto the table.

Reuben chirped up: 'Mam, these are two of mi teachers frae school.'

'Ooo, are you alright?' one of them said, looking at my face. 'You look very tired, dear.'

'Nah, there's nowt up wi' mi mam,' said Reuben. 'She's just been clubbin' in Leeds all night.'

Reuben's casual throwaway remark made it sound like a common occurrence, and I shrank from the teachers' disparaging glances.

It wasn't the first time Reuben had cast aspersions on my parenting skills. A few months earlier he'd been assigned a school project on vocations which was to culminate in a presentation at a special assembly in the church hall, to which all parents were invited. Reuben had put a lot of time and effort into the project, interviewing some of our friends, talking about their jobs. Finally, after he'd talked to the mechanics, van drivers and builders, he asked me what my job was.

'I am a shepherdess; I tend the needs of my flock. Folks 'ave shepherded sheep since lang ago. As occupations go, shepherding sheep is one o' t'oldest.'

The day of Reuben's presentation arrived and I found a seat at the back. I must admit that I wasn't concentrating hard on what the other children were saying, but when I saw a grinning Reuben get up from his seat and head for the stage, I pricked up my lugs. He spoke very well, talked of commitment, qualifications and job satisfaction. I was impressed, very much so, until . . .

'. . . And as for my mam, her profession is the oldest in the world.'

The hall erupted. Reuben looked very pleased that he'd provoked such a reaction.

Later, on the way home, he said, 'Did you like mi talk, Mam?'

'Reubs, I loved it,' I said, and I meant it. I hadn't laughed so much in a long while.

Violet is more than happy to play rough, fight, wrestle and take on all comers. But she also loves to wear pretty dresses, the more bling the better. Some of her favourite outfits come from the Appleby Horse Fair: famous for horses, but not that well-known in fashion circles . . .

The fair, which happens in June every year, is an annual gathering of travellers from all over the country. They converge on campsites, many of them in brightly painted horse-drawn vardos. The gathering is swelled by thousands of tourists and spectators. The gypsy lads bring their horses down off the Fair Hill to the River Eden, riding them bareback into the water and washing them. Then they are trotted along the Long Marton Lane, known as the 'flashing' lane, because they are 'flashing' or showing off the horses that they have for sale. The majority of horse trading is done down on the Dealers' Corner, a triangular piece of land at the bottom of the flashing lane. Here folks argue, barter and eventually agree on a price. A 'fixer' oversees the proceedings, and the deal is cemented with the slapping of hands of the buyer and seller.

It's not just horses that are bought and sold: the Fair Hill campsite is spread over acres of fields, with caravans, horse boxes, vardos and wagons interspersed with catering vans, fortune tellers, palm readers, stalls selling clothes, household goods and anything to do with horses. There are lots of rip-off luxury brands, stalls selling fake tanning lotion that turns you orange and cosmetics in almost luminous shades. It's quite a spectacle, because the girls dress up in all their finery and parade up and down. If the weather is great, it's amazing, but when it's raining and the fields are churned up with mud they don't swap the stilettos for wellies. They simply accessorize their tight bodycon dresses with plastic bags over the sky-high heels. It's these Spandex, net and diamanté-encrusted outfits that entrance Violet.

We live quite near to Appleby, and every year we try to visit the horse fair. Most years we come back with little to show for it other than a slightly lighter purse and some cheap and cheerful tack. A souvenir or two is quite acceptable to Clive: one year I found a hook-over feed trough for feeding the calves, and the year before I bought a cast-iron 'Queenie' stove, the traditional stove of the gypsy caravan. The one thing Clive forbids me from buying is a horse. At first sight I'm never much struck with the horses on the Fair Hill, but hidden among the plain, coarse, poorly bred horses there will often be one that stands out. It won't be cheap, because the owner will know that he's got a good 'un.

On one visit I was humouring Violet by looking through the clothes on the sale rails. I had no intention of buying anything, as it wasn't exactly practical gear for wearing around the farmyard. Just as I was dragging Violet away a lady, dressed from head to foot in leopard print, approached me.

'Don't ah know yer?' she said, chewing gum. 'Yer off t'telly, in't ya?'

'Yes, that's right,' I said, smiling and admiring her long neon-orange nails.

'Oi, Cheryl!' she shouted towards the white Transit van parked at the back of the stall. 'It's that family wi' all them bairns, yer knaw, they've got 'orses an' all.'

Cheryl appeared, backside first as she climbed out of the driver's side. Bigger and brasher than her friend, it was her over-sized hair rollers that caught my eye.

She came towards me, arms outstretched. 'Eeeh, I luvved that programme. 'Ow many of t'kids has ta got wi' yer today, then?' Her accent was broad Yorkshire.

'All of 'em,' I said, gesturing towards the children.

'Well, let me tell yer summat,' she said. 'I'm sixth-generation Romany, and I'm tellin' yer that yer family's gonna get bigger.'

I didn't put much store by the prediction, but smiled and thanked her.

By now Violet was in the process of pulling a mauve-and-white frilly dress, garnished with tiny flowers, off its hanger.

'Violet, leave that alone,' I said, frowning.

'Naw, naw,' said Cheryl. 'She can 'ave it. 'Ere, let me get yer a bag.' Before I had time to argue, the dress and a pink fake astra-khan gilet were being unceremoniously stuffed in.

''Ere, 'ave some babby socks,' she said, pulling out some adorned with pearlized beads and lace. 'And a bonnet an' all.' In went a baby bonnet, with a big bow on the top. 'You'll be needing 'em. Worrabout a dummy?' Before I could protest, a metallic gold-coloured dummy on a chain went in.

'It's been mighty fine meeting yer,' she said, her friend nodding, her hoop earrings brushing her shoulders.

'It's varry kind of yer,' I said. I could see Violet was delighted, clutching the plastic bag with glee.

'Say thank you, kids,' I said, and they chorused their thanks.

'Kushti bok,' they shouted back, as we set off towards the Land Rover.

'Can I put mi dress on, Mam?' said Violet, before we'd even left the field.

'When we get 'ome,' I said. The sky had greyed over, and it had begun to drizzle. As we left we saw the bow-top caravans, which were parked on the verges and roadside, had small fires smouldering next to them, with black kettles hung above on chivvies. The wooden wagons were intricately painted in deep, rich reds and greens, with swirling gilt scrolls. Tethered to fence posts nearby, coloured ponies cropped the grass. It was a timeless scene, unchanged for centuries, as long as you discounted the satellite dishes anchored to the fronts of the vans and the flickering TV sets inside.

I breathed a sigh of relief when we turned up the road to Ravenseat. Taking the children to the horse fair is not for the faint-hearted. There are so many dangers: horses being ridden bareback at high speed, young lads on sulkies racing their high-stepping trotters. Every year we've seen bystanders being hit by runaway horses and sent flying, or being caught up in the wheels of the passing carriages. Keeping an eye on my brood had tired me out.

'Mam, can I put mi new dress on?' asked Violet once we were in. I could see she wasn't going to rest until she tried it on.

'Aye, Raven will 'elp yer,' I said, putting the kettle on.

Although I was tired, the children were as lively as ever. The rain stopped and the evening was now pleasant, except for the midges. It wasn't long before the children were outside and I was sitting at the kitchen table with Clive, telling him about our day while waiting for the soup to warm up on the range.

'I'm not sure about Violet's dress,' he said. 'It's gay nylony, looks highly flammable.'

'That's as mebbe,' I said. 'But she likes it, and it didn't cost mi owt.'

I glanced up at the clock. It was soon going to be bedtime, never mind teatime. I really should be rounding them up, I thought. At that moment I heard a bloodcurdling shriek. Clive heard it too, and he's deaf – that's how loud it was. He reacted fast:

'What thi 'ell was that?' he said, putting his cup down and striding out of the kitchen door.

Seconds later Edith appeared at the door. 'Mam, Violet's 'urt, there's blood.'

I'd only got as far as the yard gate when I met Clive coming towards me carrying Violet, his hand over her forehead. Blood dribbled down the side of her face and dripped onto the farmyard floor.

The children were all crying, a cacophony of wailing. The only one who was peculiarly silent was Violet.

'What's 'appened? What's 'appened?' I shouted, running alongside Clive.

'She's fallen, it's an 'ospital job,' he said.

We got back to the kitchen. Clive sat down, his hand still firmly clamped over Violet's brow. I grabbed a tea towel.

'Let's 'ave a look,' I said. I knew from past experience that head injuries usually meant lots of blood, but as soon as he lifted his hand away, I could see that this wasn't superficial. It was a very deep cut, the gaping wound deep enough to see bone. She also had a cut to her cheek. Her blonde hair was stuck to her face with clotting blood.

I held the tea towel over the cut, trying to stem the blood flow. Violet was trying to rub her eyes, her lashes now encrusted with blood.

'You 'old the towel an' I'll ring for t'ambulance,' I said to Clive. The children gathered round Violet, their tear-stained faces a mixture of fear and pity as she lay in his arms. The new mauve-and-white frilly dress was smattered with blood and dirt.

We went through the usual litany of questions on the phone. Had she lost consciousness? No. Was there any foreign object in the wound? No. Then it was a matter of waiting and planning. It was no good me getting in the ambulance with Violet, as there would be no way of bringing her back to Ravenseat after she'd been stitched up. We pragmatically decided that I should follow the ambulance in the Land Rover, so that I'd be able to bring her home later that night. We were sure Violet was going to be fine: she just needed cleaning up and suturing, and probably a tetanus jab.

After what felt like a very long forty minutes, the ambulance pulled into the yard. Preliminary checks were done before she was loaded into it, and I explained to Violet that I would see her again when we reached the hospital. She was unconcerned, busy telling the ambulance man what had happened. Apparently she had climbed onto the straw bales in the barn and then spotted one of the chickens sitting on a clutch of eggs down a crack between the bales and the cattle barrier. In an attempt to get a better view she'd leapt the gap, but miscalculated the distance and fallen head first, hitting the concrete. I couldn't believe that after spending the day chaperoning the children around the Appleby Fair and avoiding all the dangers there, an accident had happened under my nose in our own farmyard.

It wasn't easy keeping up with the ambulance as it sped along the country roads but eventually, at around 9 p.m., we pulled up at the Darlington Hospital Accident and Emergency department. Violet was wheeled in, clutching her dolly and complaining bitterly about her spoilt dress.

'Never worry,' I said. 'It'll wash.'

We waited a while to see the doctor. Removing the gauze bandage and dressing the ambulance man had put on, he squinted at the gash in Violet's head.

'Mmmm, it's a deep cut, clean but very deep,' he said, pulling the skin either side apart. I held Violet's hand in mine: she winced, frowned, but didn't cry.

'Right,' he said. 'She needs plastic surgery, so I need you to have her at Durham University hospital first thing tomorrow morning. Do not let her eat anything from midnight.'

I hadn't expected this. What was I going to do? Durham is seventy miles from Ravenseat, so it seemed pointless going all the way home, then setting off in only a few hours to get to Durham. Plus Violet was complaining about being hungry, and there wasn't much time left before midnight to feed her. I rang Clive, told him what had happened, and we both came to the same conclusion: I should call at a fast-food outlet, then find a hotel for the night.

The fast-food bit went without a hitch. Then we went in search of a hotel, not expecting it to be difficult. I started with the more reputable hotels, then, becoming increasingly desperate, resorted to less salubrious-looking places. Each time it was: 'No room at the inn.'

I gave up on Darlington and set off towards Durham, spotting a seedy-looking motel at the roadside. Pulling up outside, I felt sure that we'd get a bed for the night. I walked into a dingy reception clutching Violet's hand. Behind a metal shutter sat a bored-looking receptionist, texting on her mobile phone.

'Any rooms for tonight?' I said.

'No,' she said, not even looking up from her phone.

By now it was nearly 1 a.m. and I was giving up hope of finding anywhere, and resigned myself to driving to Durham hospital and sleeping in the Land Rover in the car park: not a prospect that I relished. We set off again, Violet now dozing in her seat. I was in unfamiliar territory, but thankfully there was little traffic so I could drive slowly, watching the road signs. We were soon on the one-way system in Durham city, and I saw a well-lit hotel, blue fairy lights strung around the door and a sign saying 'Champagne Bar'. A couple stood outside leaning against the wall, taking the night air. Doing an emergency stop, I decided that this was my last-ditch attempt to get a room for the night.

'C'mon, Violet,' I said softly. She never stirred as I picked her up. With me carrying her in my arms, we must have been quite a sight as we went through the pristine white foyer to the reception desk: Violet's dress was stained with congealed blood, her tousled hair straggled out from beneath a giant bandage, I was still wearing the clothes that I'd worn for the fair: wellies, leggings and a camouflage jacket. I looked around at the smartly dressed couples sitting at small tables drinking cocktails, having hushed conversations with faint strains of blues music playing in the background. I was expecting raised eyebrows at the scruffy gatecrashers who had invaded the party, but rather than turn tail, I decided to give it one last go.

'Hello, I just wondered if you had a room available for tonight, please,' I said, probably sounding defeated and tired.

The well-groomed young lady behind the desk spoke with a thick foreign accent. 'Oh my goodness,' she said looking at Violet, her head flopped back over the crook of my arm. 'What 'az 'appened?'

I told her our tale of woe. She listened intently.

'I just need a room, no breakfast,' I said. Violet was still fast asleep.

'Do not vorry about anyzing,' she said. 'I 'ave a room for you . . . and I'm going to upgrade you to one of our business rooms.'

She handed me the keys.

'I need to park my car,' I said. I'd left the Land Rover obstructing the drive, not really expecting to be staying.

'I tell you vot,' she said, 'I weel votch ze baby while you do zat.'

I didn't sleep well that night, but Violet slept soundly. We were up with the lark and before we left for the hospital treated ourselves to a very deep, hot bath, then wrapped ourselves in the deliciously soft towelling robes. We could have been on holiday – apart from the bandage round Violet's head. We set off for the hospital, expecting that Violet would be patched up and we'd soon be heading home.

When we arrived she was assessed by a doctor, but then we were told that she needed an operation and the plastic surgeon was unable to do it until the next day – in the afternoon if we were lucky. Violet was happy, and had another clean dressing on her head, so we decided to get out of the hospital and go sightseeing, as well as picking up a few essentials for our unexpectedly prolonged stay. Violet, although now clean, was still wearing her bloodstained dress, but we managed to conceal the worst of it under the pink fake-fur gilet that had been left in the Land Rover, part of Violet's treasure trove from the trip to Appleby. We looked around the cathedral, wandered down the main street, then shared an ice-cream sundae in a parlour near the riverside. It would have been very pleasant if it hadn't have been for the hole in Violet's head.

We were back in our room at the hospital by teatime, and spent an uncomfortably hot night there. The next day really dragged, because all we could do was sit around waiting. Violet was given her pre-med, and in order to take my mind off things I decided to use my time wisely and write an article that I'd been asked to do for a parenting magazine. It was about relaxed, hands-off parenting: letting children climb, run and play without worrying. The irony was not lost on me.

Violet had the operation and we were back at Ravenseat by bedtime on the Monday, after our unscheduled two-day mini-break in Durham. Our trip to the horse fair seemed to be a long time ago, so much had happened. Violet has two small scars which will fade in time. I expect there will be further accidents from time to time at Ravenseat: children will be children.

I brought horses back to Ravenseat when I moved in with Clive. Horses were at one time an integral part of farming in the Dales. Used for shepherding, now replaced by a quad bike, used to work the land, now replaced by a tractor . . . their only role now is for leisure and companionship. Talking to older farmers, there is a definite split in the attitude towards the horse power of yester-year. Some have fond memories of the ponies, their names and their idiosyncrasies; others were pleased when mechanization took over. My old friend Jimmy told me about a pony being yoked to a trap laden with all manner of crockery and china to be taken to the chapel at Keld. The pony jibbed, stood rooted to the spot and then decided that the way forward was actually backwards. Unfortunately there was a steep drop at one side of the road and the trap, still harnessed to the pony, was left dangling over the edge, with all the plates and cups and saucers tumbling down the embankment and smashing.

Tot, our elderly neighbour, remembered that during the war years an attempt was made to plough some of the land in the upper dale. The earth there is so heavy that when the horse took the strain of the single-furrow plough, it stubbornly refused to move forward. The solution, perhaps a little unconventional even by the standards of the time, was to light a small fire under it. It soon moved.

Whether it was tales of Jimmy's pony ruining afternoon tea at the chapel, or our friend Frankie's horse saving the day by bringing him home safely from the pub after a long session, somehow it seemed right that horses should make a comeback at Ravenseat. We have over the years used them for shepherding, and even once for pulling a bogged quad bike out from its watery grave high on the moor. I've always loved horses, going right back to when I scrimped the money from my Saturday job to pay for riding lessons as a teenager. Clive, on the other hand, has an ambivalent relationship with horses: he appreciates them as animals, but he has no desire to ride them. I can only remember him climbing on a horse's back once, and I don't think he or the horse, Meg, enjoyed it.

'Mand, it's moving,' he said, shuffling uncomfortably in the saddle.

'*It* is called Meg,' I said. 'And she's supposed to move, that's the idea of going for a ride.'

Then he tried to get off, swinging his leg over Meg's broad back. It was no smooth dismount, as his dodgy hip meant that he got stuck in the process.

'That's the first time I've ever known thee to 'ave problems gettin' thi leg over,' I laughed.

Clive is wary of horses and they, too, treat him with an air of suspicion. He is the complete opposite of a horse whisperer, having a natural ability to turn any perfectly sane, good-natured horse into a complete loon.

The children love, understand and show great respect for the horses, and I believe that the horses reciprocate their feelings. The older children ride them down from the moor bareback, through the heather and bracken, crisscrossing the little rivulets and following the sheep trods. Spending their summers turned out at the moor makes the horses sure-footed and unlikely to stumble. The younger ones trundle around the farmyard on our Shetland pony Little Joe, who is as saintly an animal as you could wish for. Bandaged from head to hoof for a game of doctors; his mane plaited and replaited and decorated with ribbons; even pulling skiers through the snow: he never demurs. The children have fun and Little Joe obligingly trots around with them, so I figure that he must enjoy the attention or he would head for the hills. He even (unknowingly) helped me with Violet's toilet training. I'd never had much trouble when it came to getting the children toilet trained; I simply let them roam around the farmyard or field without nappies until they picked up on what was happening. Violet was more stubborn, and to persuade her to use the toilet I resorted to bribery.

'If you use the toilet then I'll take you for a ride on Little Joe,' I'd promise.

I had Little Joe on permanent standby in the garth. My plan worked a treat. I'd hear a little voice shouting from the downstairs toilet: 'Maaam, can I go for a ride now?'

The shouts became more and more frequent. Violet developed superb bowel control, and what should have merited one pony ride could be spun out to three.

We have another Shetland pony, Folly, who the children bought for thirty guineas at the Cowper Day horse fair. They were on the field gate, helping to collect the parking money, pocketing all the loose change for themselves. When all the cars and horse lorries were parked they sloped off into the auction and bought themselves a horse. I admit that I had been quite vague and had said, 'Go an' get yerselves summat,' meaning perhaps a new bridle or lead rope. ''Ave a look around the stalls, you'll see summat, I'm sure.' And they certainly did see summat: a very small pony, which came home with us crammed into the back of the Land Rover.

The children can ride Folly, but she's not as amenable as Joe. She tolerates it all for a while, then turns into a bucking bronco. Edith rode her the other day, and she was fine. Then Miles decided that he should go for a ride too: putting his hard hat on, horse and rider rode sedately down the yard, past the farmhouse, picnic benches, across the river, and into the bottom field. The other children were already playing there. I don't know what happened, but suddenly Folly reared. Miles was launched into the air and landed in a heap on the grass, still holding the reins, as every good horseman should. He was fine . . . until Folly stood on him. Swinging around, she put her tiny, poky little hoof right on his crown jewels. He yelped, loudly. All I could see from my vantage point in the farmyard was the children doing some kind of war dance around a small skewbald pony, and a prostrate Miles. There wasn't much in the way of sympathy for Miles, as the kids just found it very funny. Like all good riders he was not put off, and he was soon back in the saddle – only this time, he wore Reuben's cricket box as well as a helmet.

As a rule the coloured horses, the gypsy vanners that we keep, are good-natured. They have the odd altercation, but they generally try to please. We do the basics when riding, nothing fancy:

left, right, stop, start, faster and slower. Over the years we have accumulated everything you could ever need for a horse: bits of every size, bridles, saddles and rugs. In one barn loft we found harnesses that belonged to horses whose names I will never know. The straw-stuffed collars are bursting at the seams where the leather has dried and split, there are patches in the leather-work which has been stitched by hand, there is an extra hole in a strap made here and there, and horse hairs still cling to the fibres and poke out from the jute underside of the cart saddle. Sentimentality stops me from doing the sensible thing and having a clear-out. Boxer's old saddle gathers dust in the porch, Stan's Witney rug is still folded in the blanket box. I worry that I may in later life become a hoarder, surrounded by a sea of tack from long-dead horses.

Meg came to us almost accidentally after we spotted her teth-ered by the side of the road at a small gypsy encampment, at the time of the Appleby Fair. Clive may not be particularly fond of horses, but he certainly knows a good sort, and Meg fitted all the right criteria.

'If you've gotta 'ave a 'orse, you might as well 'ave a good 'un. It costs as much to feed a good 'un as a bad 'un,' he says. It's a mantra that he repeats for all livestock.

Because we run the car park at the Cowper Day horse sale at Kirkby Stephen, we get to hear where the good stallions are. It was through our connections that we heard about the Black Stallion, a beautiful, much-coveted stallion that was making a name for himself in horse-dealing circles. We were lucky enough to be offered the chance to have one of our mares run with him for a month, with the intention of getting her in foal. Clive took Meg's daughter Josie in the cattle trailer to Blaydon, in Tyne and Wear, where the Black Stallion was kept. When he got back he said that the horse was everything that we'd heard: small, compact, plenty of bone and of course jet black with a white blaze on its nose.

'Lost i' feather, reet from back of 'is knee to t'floor, a helluva thing, yer could put 'is muzzle in a pint pot,' was his verdict.

Clive had been taken on a guided tour of the stables. They were nothing fancy, mostly built from sheets of corrugated iron. In the ramshackle yard was a static caravan in which the owner lived.

'I see'd all sorts,' said Clive. 'Some hellish 'osses, some reet good 'uns. They offered to sell me yan,' he said, though I could see that he'd come home with an empty trailer, so he hadn't bought it.

'It's a good-looking mare. She lost 'er foal and she's running wi't stallion. She's not in good fettle, but we can 'ave 'er for a thousand quid.'

I hadn't seen the mare, but the fact that Clive had taken a shine to her meant that she must have been something a bit special. 'Well, if thi likes 'er, then thi'd better bring her yam when yer pick Josie up,' I said.

A month later he went back to Blaydon. The children and I excitedly waited for him to get home with, we hoped, an in-foal Josie and a new addition to our equine family. We heard the trailer rattle as it came over the cattle grid.

'They're here, Dad's back!' Raven shouted, hopping around in excitement in the middle of the yard. A welcoming committee assembled by the yard gate to greet Clive – or, more particularly, the new horse. We were pleased to have Josie back, but when the trailer door was let down, all eyes were on the new arrival.

Josie knew she was home. Sweated up from the long journey, she clattered down the trailer door and then stood stock-still in the yard, raising her head, her nostrils flared. She whinnied, as if to announce her arrival to anyone who would listen. In the distance, one of the other horses answered.

'Take Josie through to t'sheep pens an' let 'er through t'moor gate,' said Clive to Raven, handing her the lead rope. She set off at a jog, turning round to shout:

'Don't get the new 'orse out 'til I'm back.'

Within minutes she was back, swinging the lead rope round her head like a lasso.

Clive went into the trailer, pulled the catch and swung the middle door open. An off-white, long, whiskery face with a twirl of moustache looked back at us. Clive attached the lead rope to her head collar and walked her down the ramp. For someone who really wasn't so keen on horses, he seemed to have taken a shine to this one.

'Here yer are, owd lass, this is yer new 'ome,' he said, patting her neck. The children said nothing. Raven went over to the mare to have a closer inspection.

''Ow old is she?' she said.

'She's a bit of age,' said Clive.

And an old lass she certainly was – probably around twenty, we thought.

'What's 'er name?' asked Raven, as she gently stroked the mare's velveteen muzzle.

'Dunno, erm . . .' Clive said.

'Queenie,' I said. 'She's called Queenie.' It suited her.

'Put 'er in t'top garth when thoo's finished lookin' 'er over,' said Clive. 'I'm gonna tek the trailer off.'

Off he went, leaving us clustered around the new horse, Queenie. The anticipation and excitement of the morning had dissipated. It wasn't necessarily disappointment, but a reaction to the feeling of weariness and innate sadness that seemed to emanate from her. She had 'rain scald', weeping sores on her back where her skin had become infected from being constantly damp. She was white, or at least she would have been, but her coat was very dirty and felt greasy to the touch. Her tangled mane was thinning and lay lank against her neck, parts of the mane having been scrubbed away at the roots so the skin along her crest had thickened and scabbed over. She was slack-backed and big-bellied, but whether the former was entirely due to her age or because she had been ridden as a youngster, we would never know. The latter was certainly the result of being used as a brood mare for many years – and I should know about that!

'Mam, d'yer think she's broken to ride?' asked Raven, studying Queenie hard.

'I just dunno, I wouldn't 'ave thought so,' I said. 'But I imagine that she'll 'ave been broken to harness.' For she was a cart horse if ever I'd seen one, broad-chested and sturdy-legged. Whatever had happened to her in the past, she would never be ridden or driven again by anybody, I vowed to myself.

'We need to ring t'farrier,' said Raven.

Oh yes, the hooves. They were in a bad state, curling up like Turkish slippers, forcing her down on her heels, which was putting pressure on her tendons and causing her to move with an awkward and uncomfortable gait. Her nearside back hoof had split right up towards the coronet, and with each step that she took the split would gape a little wider. I tried to pick up her legs, but it was a step too far: she wasn't having that. She didn't lash out with any real malice, but in her watery blue eyes I could see real fear. It is impossible to describe what a kind eye actually is; you just know if you see one staring back at you. I sensed that Queenie had a kind and gentle temperament, but she needed time to adjust to her new life, and forget the fear of her past one.

I rang Steve, the farrier, explained the situation, and warned him about the work needed and that there was a good chance she would be bad to deal with. It didn't faze him, and he agreed to call by in a couple of days, when hopefully she would be more settled.

Raven and I decided that a bath in the river would be the perfect starting place for Queenie's new regime, and once we had got her clean we could set about the task of treating her wounds. Hopefully with the application of antiseptic green salve we'd get her skin ailment under control, if not completely cured. But that was all to be done tomorrow: today it was about turning Queenie into the top garth, letting her relax and get her bearings.

We woke the following morning and looked out from the bedroom window and across into the top garth, where the early-morning sun was shining on a sea of yellow kingcups. Queenie lay amongst the flowers, perfectly still.

'D'yer think she's alright?' said Clive.

'I dunno,' I said. Clive pulled down the sash and, leaning out, gave a loud, long whistle that cut through the stillness. Queenie never stirred. Out I went in my nightdress, making my way across the yard and then through the dewy grass. Swallows dived and swooped as my footsteps wakened the insects and sent them skywards into the path of the hungry birds. Climbing the gate, I hurried towards Queenie. I was getting worried now, and I could see Clive was still hanging out of the bedroom window, waiting for news. I got right up to her, her bloated belly looking even bigger when she was lying down, her legs outstretched so I could see the full state of the overgrown hooves. When I was so close that I was within touching distance, I heard what sounded like snoring. I spoke softly: 'Queenie.'

She blinked and then lifted her head, and I looked towards Clive and gave him the thumbs-up sign. She clambered to her feet, then shook herself, sending loose hair and skin flakes into the air. Finally she snorted, as if disgusted that I'd roused her from her slumbers.

Horses can lock their legs and sleep standing, but they can also sleep lying down, usually when they are in a herd with another horse on lookout. They are a prey animal, and lying down leaves them vulnerable to would-be predators. It was a good sign that Queenie felt relaxed enough to have been caught napping in the field. I had a feeling that the old lady was going to fit in to life at Ravenseat just nicely.

Her long-overdue bath went well. She actually seemed to enjoy being soaped up and even tolerated us daubing on the salve, which must have stung. The farrier's visit proved more problematic: dressing the front hooves took a while, and she showed her displeasure by nipping at Steve's backside. But when he got to her back hooves, the real trouble started. Running his hand down the outside of her flanks she tolerated, but the minute he got further down towards the fetlocks she'd let fly with a kick. I clasped her lead rope tightly, keeping her staring dead ahead, her ears twitching as I whispered words of encouragement: 'C'mon, mi lass, it's for yer own good, yer know.'

My words were wasted. She let fly properly, her leg pumping back and forth like a piston, and it was only a matter of time before one of her hooves made contact with Steve. All credit to him, he never lost his temper, he never elbowed her in the guts or swore, but eventually he let go of her leg and took a step back.

'It's naw good, we're gonna 'ave to twitch 'er,' he said, taking the cigarette that had hung from the corner of his mouth for the duration of the encounter, and holding it momentarily between his thumb and forefinger. He stood looking at Queenie for a second, then after a final drag on the cigarette he dropped it to the floor and stubbed it out under his boot.

'Yer knaw 'ow this works, don't yer?' he said, returning from his estate car carrying a cosh with a short loop of rope at one end. I winced. It was a twitch, and I knew exactly how this primitive-looking implement worked. After gently stroking Queenie's nose he looped the rope round her upper lip, and then twisted it tight. It's an old-fashioned method that calms a horse down almost instantly, releasing endorphins that make the animal sleepy and compliant. The only other choice would have been to postpone the whole procedure until we got a sedative from the vet's. I wasn't enthralled with the idea of injecting Queenie, who I suspected was possibly in foal, with any kind of drugs. The twitch was the lesser of two evils.

Holding the wooden twitch steady and keeping the rope right on her lip, Steve soon had Queenie dozy, standing quietly almost in a trance. He was able to dress her rear hooves, trim them back and rasp until the angle of her pasterns was corrected. It was truly marvellous to see her afterwards standing correctly. Once the twitch was released she snapped back out of her hypnotic trance and showed us a clean pair of heels as she was turned back into the garth.

Slowly, as the weeks went by, Queenie changed. The painstaking application of salve did its job, coupled with brushes and baths with Dermoline, and she began to blossom. She found a new lease of life, galloping to the top of the field and back again just for the sheer pleasure of it. She was biddable, got to know her name and began to trust us. By the time winter came she was happy to be rugged up, and the smaller children could lead her in and out of the stables to water her. During these winter months it became obvious that, like Josie, Queenie was definitely in foal, and we had to loosen her rug by letting out more of the surcingle to accommodate her swelling tummy. After tea sometimes we'd take Queenie out for a walk on the lead rope to graze in the garden. Violet and Edith would stand close beside her, their ears pressed against her sides, listening for the foal: they were delighted if they saw or felt a kick. We watched her carefully, looking for the tiny droplets of waxy milk on her udder that would signal the foal's arrival within twenty-four hours. It was a cold March morning when she gave birth to a filly, beige and white, strong and healthy. We decided to call her Princess. I was worried about Queenie having enough milk to feed her, knowing that she'd lost her last foal. I reckoned that it would make sense to help things along and bottle-feed Princess with a mare's milk replacer. I baulked at the price – a hundred pounds for a bag – but it was better to be safe than sorry.

Princess was an ugly duckling: she had the sweet and innocent look that all foals have, but she wasn't strikingly beautiful. This was very fortunate, for her and for us, because one day her 'rightful' owner came to Ravenseat to reclaim her and Queenie.

It was a glorious afternoon, with the older children at school and the younger ones busy playing around in the yard. I heard the noise of a vehicle and the clatter of a trailer coming up into the farmyard, and immediately stopped what I was doing and went to see who it was. A battered flatbed Transit van and horse trailer had parked outside the farmhouse door, and a man got out of the driver's side.

'Hello,' I said. I thought he was a scrap man, and I was wondering whether I dared 'donate' some of Reuben's paraphernalia that was littering the farm: old broken bicycles, springs, wheels and other junk.

'Are you Amanda?' he said.

'Yes, what can I do for yer?'

'You've got 'orses, 'aven't yer?'

'Erm, yes,' I said, feeling a bit uncomfortable and furtively looking sideways to see where I actually had the horses. Meg, Josie, Folly and Little Joe were out of sight, probably further up the moor bottom; Queenie and little Princess were quietly grazing in the top garth. 'Why do you ask?'

By this time, Clive had appeared on the scene. 'Alright, lads,' said Clive, including the other man in the passenger seat in his greeting.

'Aye.' The conversation was now directed at Clive. 'Did you buy a white mare off Black Jonny at Blaydon?'

I had a nasty feeling that this wasn't just a social call. I looked hard at our visitor, standing with his hands in his pockets, greasy hair fastened back into a ponytail, unshaven, with thin lips and piercing eyes. There was a mean look about him. I glanced at his passenger, who was fiddling with his mobile phone. He was built like the proverbial brick outhouse. He slowly returned my gaze, his face expressionless. I was right to be worried.

'Aye, we did,' said Clive. I picked up a faint nervousness in his voice, and hoped it wasn't noticeable to these two thugs.

'Well, it weren't 'is to sell,' the first man said. 'It were mi boss's, an' 'e wants 'er back.'

They said that their boss had taken seven mares, including Queenie, to be served by the Black Stallion. Queenie was in her foal heat, twelve days after foaling. She lost the foal she had at foot for reasons they didn't know, and when the boss came to pick his mares up a few weeks later, one of them was missing: Queenie.

I guess if you deal in these circles, then sooner or later you're going to run into trouble, and this felt like trouble with a capital T.

I wasn't quite sure how Clive was going to play this, but I knew that there was no way I was going to be parting with

Queenie and Princess. Not only had I forked out a thousand pounds for her, we had lavished her with love and care for nearly a year. Nope, I didn't care that their faces looked like mugshots off *Crimewatch*: they were not taking Queenie away with them.

'Now c'mon, chaps, let's be reasonable 'ere,' said Clive. (I severely doubted whether these guys did 'reasonable'.) 'T'old lass is over there,' he went on, nodding towards the contented mare and foal grazing in the field. 'She's knackered, been at death's door, we're 'avin to feed her foal for 'er. An' foal's nowt I'll tell yer. It's a filly, but not a good 'un.'

I bit my tongue. I didn't like Clive saying such things about Queenie and Princess, but I knew what he was trying to do.

'I'll be t'judge o' that,' said my unwanted visitor. 'C'mon, Billy, let's ga an' ave a look.'

Billy lumbered out of the Transit, shoving his phone in his back pocket and pulling his vest down over his hairy, protruding beer gut. I looked at Clive, and he looked back at me. As the pair set off towards the gate, I grabbed Clive's arm, pulling him back.

'Yer not really gonna let 'em take 'er, are yer?' I whispered.

'Eh?' he said.

'They're not 'avin mi 'orse, not whatever,' I said defiantly.

'I just don't want any trouble. An' you're in no state to argue,' he said, for I too had a protruding belly, but for entirely different reasons to Billy's.

Billy and his friend were talking and looking up the field. Queenie was now standing squarely on top of the hill, staring back, oblivious to the fact that her circumstances could be about to change – and not for the better. Princess was nestled in at her side. Clive and I went over.

'My missus 'ere says that she'd rather not part with 'er 'orse,' he said.

I was getting angry now; maybe my pregnancy hormones were causing me to be a little more overwrought than usual, or maybe it was just the tougher side of my upbringing in Huddersfield. I flew mad at them. I swore, a stream of words coming out of me that I don't normally use. I told them in no uncertain terms that they must leave and they'd not be taking my horse with them.

Clive looked embarrassed as I carried on ranting.

'Look lads,' he interjected. 'You'll lose t'foal if yer tek 'er and don't bottle-feed 'er and yer mare'll nivver breed agin, yer'll end up wi' nowt.'

He was right.

'An' you need to calm down an' all,' he said to a red-faced me. 'Or you'll end up foalin'.'

He turned back to the two men.

'What if I load yer up wi' a few bales o' hay for t'winter, and let's be calling this matter closed.'

After a quick discussion an agreement was reached: a trailer full of hay would calm the waters, and they would report back to their employer that Queenie was at death's door. No hands were shaken – that's what gentlemen do, and these chaps didn't fit into that category.

I stood with my hands on my hips, my eyes narrowed and a scowl on my face while they backed their trailer up to the barn. Clive threw the bales down from the mew (stack) and loaded them up. Nothing more was said, and it was only as I watched them trundle down our road that I began to feel shaky.

'Yer dozy mare, you could 'ave gotten me a good kickin' there,' Clive said, frowning at me. Then he smiled: 'Even I were scared o' yer for a moment there.'

The funny thing is that Princess has turned out to be a very good horse, really blossoming, with the beige turning black, her body filling out, with a full, thick tail and heavily set. She has all of Queenie's attributes. Her only real failing is her ability to get dirty. Not all horses are the same when it comes to the call of nature, which you soon discover when you have stables to muck out. Josie and Queenie would do their droppings all in the same place, around the side of the stable, nice and easy to shovel out and never, ever would they lie in the poop. Meg and Princess were a different kettle of fish, both making a terrible mess in their stables. But Princess would also lie and roll in it. Every night I had to bank up the walls of the stable with straw to prevent Princess getting cast, stuck on her back during the night.

One morning I got Princess out of the stable in a terrible state. I said to Clive, 'I'm gonna pressure-wash Princess.'

'Yer can't pressure-wash an 'orse,' Clive said.

'Well, the sheep don't mind their faces an' legs being washed with it. They prefer it to the cold-water hosepipe, so I'm gonna give it a go,' I said.

I turned the water to warm, turned the pressure right down – and she loved it. She looked terrific, but of course it didn't last long, and she was as dirty as ever within a couple of days.

There have been times when the little ones have been playing in muddy places, with the mud slowly migrating up the insides of their leggings from their wellies until their whole lower half is covered, when the easiest solution is to lay the clothes out in the yard and blast the muck off with the pressure washer, one side and then the other. I have, so far, resisted the urge to do this with the children inside, but who knows, one day . . .

Not long after this incident, Josie foaled prematurely. The weather had improved, so we had turned her to the moor bottom, bringing her into the stable every night. One day we saw her looking over the wall into the High Bobby Dale, taking great interest in Queenie's foal. We didn't realize the significance until teatime, when there was no sign of Josie at the moor gate. She usually made her way back here knowing that her warm, cosy stable was waiting, with a bowl of soaked sugar beet and a full hay net. The weather was fine and I had no worries about her spending the night outside, but it did seem a little odd, as the other horses were all in the vicinity.

'She'll 'ave broken away from t'others to foal,' said Clive.

'She's not due yet,' I said. 'I'll go on t'bike an' see if I can see 'er.'

Up the beck and then climbing out on the top of Robert's Seat, I looked down upon Ravenseat, scanning the hollows and ghylls for any sign of Josie. The only places that I couldn't see were the steep edges below me, and in order to survey them I had to go right up to the High Force waterfall and the Graining Scars and cross onto the other side of the valley. Traversing the steep slopes carefully, about two miles from home I found evidence: a fresh placenta, still slightly warm to the touch. At least I now knew that she'd foaled and that she must be in close proximity as obviously she couldn't have travelled far with a newborn foal in tow. Switching the bike's engine off, I looked around, but all I saw were sheep moving here and there, and rabbits darting in and out of the bracken patches, occasionally raising a pheasant that chattered as it took to the sky. On a hillock some distance below was a circular stone stell, where once sheep would have been gathered to shelter from a storm – some stells even having adjoining little stone buildings where a shepherd, too, could sit out bad weather.

In this stell I could see Josie. I was amazed she had managed to squeeze through the small wicket gate, but there she was, right in the middle. I was looking down on her skewbald, table-top back, and occasionally she'd put her head up, look around, then duck down again out of sight. I assumed she was tending to her foal. Off I went, down the hill, on foot now. When I reached the wall of the stell I peeped over the top. Josie stood guard over a beautiful bay foal with black points which was sitting in front of her, long legs neatly folded under, a slender neck and a contented expression on her neat little face. She was certainly before her time, premature, a lot smaller than she should have been. The giveaway was her dome-shaped head, common in premature animals. Slowly and quietly I made my way into the stell and towards the foal, who now got to her feet falteringly. Josie, completely ignoring me, nickered quietly to her, all her attention on her newborn. The foal, a filly, had suckled, and Josie, although a first-time mother, was clearly doing well.

I knew that it would be impossible to walk them back to the farm: it was much too far for the foal, and over difficult terrain. After taking a picture of mother and daughter to show Clive and the children, I set off back uphill to the bike. My advanced state of pregnancy meant that it tired me plenty, and I was puffing and panting. I decided that the best course of action would be to take Raven with me on the bike the following day and let her slowly walk the mare back to Ravenseat while the foal (which we named Della) followed on behind.

It took two days to get the pair back home: when the foal got tired we'd stop, then return a little later and walk a bit further. By the time we were all back in the yard, a welcoming committee had assembled.

'Ooh, now she's a bit smart . . .' said Miles.

'She's gotten a white foot,' said Edith. 'There's a poem about that, but I can't remember it.'

'One white hoof, buy it; two, try it; three, suspect it; four, reject it!' I said, smiling. 'That means Della is a keeper.'

'Nah,' said Clive. 'One white foot, keep her not a day; two white feet, send her far away; three white feet, sell her to a friend; four white feet, keep her to the end. Looks like we're gonna sell 'er,' he said smugly.

'A good 'orse is nivver a bad colour,' said Raven, who can be annoyingly wise beyond her years sometimes.

At one point we almost lost Della thanks to Keith the Beef's destructive behaviour. One day Raven reported that the pony was missing. A full search party of children set off, and eventually they found her laid out in a seave bed, one of her hind legs caught up in a wire fence that Keith had brought down. Quick-thinking Raven sat on her neck talking to her and calming her, preventing her from struggling, trying to get up and injuring herself, while Reuben ran back to the farmhouse for wire cutters.

'Bloody bull,' I muttered as I cut through the wire, with Josie standing close watching us.

So that is how we came to have seven horses – well over quota, as Clive tells me to limit them to two. The number was soon to increase when we took pity on a small, undernourished colt yearling that had been abandoned in a field belonging to one of our friends. We called him Bert. He came to us as a ragged, pot-bellied, timid little chap and left us a year later as a lithe, fit and frisky fellow – leaving his mark in more ways than one. After an initial wormer dose and a few weeks of intensive feeding, we turned him loose into one of the allotments, the rougher grazing suiting him well, and he fairly blossomed. Every day we'd look across from the bottom of the farmyard and see him, a solitary figure, grazing amongst the tussocks contentedly. Then one day there was no sign of him. It didn't take long to locate him, as the commotion could be heard from the kitchen door, a cacophony of squealing and whinnying. Bert had either found a gap or made a gap in the wall and had decided to investigate what lay beyond the confines of the allotment. The old adage about the grass being greener on the other side was true for Bert, because he found some new companions: Queenie, Josie and the foals, old Meg, Little Joe and Folly.

It wasn't an easy job persuading Bert to leave his new friends, but eventually we corralled him and put him in a stable until we could wall the gap and make the allotment secure again. The whole episode was forgotten until about six months later, when we began to notice that both Meg and Queenie were looking rather rotund. Bert's foray into their territory was the only possible explanation. A visit from the vet confirmed what we already suspected: Meg and Queenie, both of whom should have known better, were in the family way.

'I din't think that Bert 'ad it in 'im,' I said to Clive.

Meg hadn't bred for six years: Josie was her last foal. She had run with stallions since then but had never got in foal, so I had decided that she was past it. Queenie, meanwhile, had a foal at foot but she, too, was long in the tooth and was going to be retired from breeding. Somehow Bert had impregnated them both in the half a day that he'd spent with them.

'I don't know how he physically managed it,' I said to Clive. 'No way is he tall enough to . . . yer know . . .'

'Where there's a will there's a way,' said Clive, and he reckons to know these things. 'He'll 'ave cornered 'em both on an 'ill end somewhere.'

You'd have thought that the two ladies-in-waiting would have got on well together, but that was not to be. Meg hated Queenie, and the feeling was mutual. When I first introduced Queenie to the herd I fully expected bickering and in-fighting, a power struggle of sorts until the pecking order was established. But what actually happened was that Meg, the sweet and gentle mare that she was, morphed into a malicious, biting, kicking, downright nasty piece. They'd both pull terrible faces, baring teeth, ears flattened back to their heads. All-out war would be declared. Meg won every time, because she was the matriarch, the leader. Poor Queenie was completely ostracized. This could have caused endless problems if they had been confined to a small field, but fortunately where they grazed in the allotments and at the moor there was plenty of room for both of them to go about their daily business without coming into close contact with each other.

Meg foaled first, a colt we called Spirit. She milked him well, and the mischievous little chap has always been a character. The children rode Meg bareback around the farm, and Spirit followed along, occasionally nipping at the children's heels when Meg stopped for a bite of grass. When Sidney took a scoop full of pony nuts into the field for Meg, she hoovered them up before Spirit ever got a look-in. But he would take great delight in picking the empty scoop up in his teeth and tossing it into the air, over and over again. As Sidney tried to retrieve the scoop to take it back to the feed store, Spirit moved it further from his grasp.

We knew that Queenie, too, would foal at roughly the same time, so we were on the lookout. We put her into the Low Bobby Dale, right in front of the farmhouse door so we could monitor her. Sure enough, the following Sunday we saw the telltale sign of an imminent foaling: tiny drops of waxy colostrum on her teats. It was a glorious June evening and the air was warm, just a gentle breeze keeping the midges at bay. We ate our tea outside on a picnic bench, watching Queenie over in the field, her tail swishing as she traipsed around, never settling. There could have been no finer time or place for her to have brought new life into the world. We lingered outside until dusk, just talking.

'When's t'foal gonna come?' asked Edith, perched with Violet on the wall, watching Queenie.

'Not long,' I said.

'Likely tonight,' said Clive.

Eventually I chased the children off to their beds, ignoring their arguments that they weren't tired.

'School in t'morning an' mebbe a new foal too,' I said.

When all was quiet, everyone in their cots, in just my night-dress and bare feet I took a stroll across the field to see Queenie.

'A foal for you in t'morning,' I said. 'And I promise it'll be yer last.'

It had never been the plan to get her in foal again, but I saw it as a blessing rather than a misfortune. I left her to it.

The bedroom curtains are never closed at Ravenseat. The sun rises in the east over the Close Hills, the early-morning shafts of light radiating the long length of the arched staircase window and into the house, rousing us all. I padded across to the bedroom window, rubbing my eyes. Queenie stood in the corner of the field, her head lowered.

'No foal,' I said to Clive. I was worried.

'Really? Well, I think that we need to be 'avin a look,' he said, rapidly pulling on his clothes. We were both dressed and downstairs in minutes.

'Fetch a halter,' he said. 'We need a hod of 'er.'

We hurried across to her.

'Shove 'er towards t'wall, I don't wanna get kicked,' he said. 'Yer gonna 'ave to hold her tight whilst I feel what's goin' on.'

I slipped the halter over her head; she never moved. We were both well versed in complications of labour, but only in bovine and ovine patients – never before had we had any trouble with an equine labour. I held the end of Queenie's tail aside, and the halter too, whilst Clive investigated.

'Nay, things ain't as they should be,' said Clive, frowning. 'It's comin' wrang. It's definitely backwards – I can feel a tail.'

'Oh, God,' I said. I knew that this was a bad thing where horses were concerned. A breech presentation of a calf or lamb was manageable, but an unborn foal has particularly long legs, making for a difficult delivery. Unborn lambs and calves can also tolerate being stuck for a while, but an unborn foal quickly succumbs to stress.

'We need a vet . . . and sharpish,' Clive said.

'I'll ring 'em, it's out o' hours but there'll be a vet on call,' I said, loosening the halter, slipping it off and putting it on the wall top. 'There's nowt we can do until t'vet gets 'ere.'

Queenie was still standing dejectedly by the wall as we both hurried back to the farmhouse. Sarah, the vet, said she would be with us in half an hour. In the meantime the older children, dressed ready for school, were sworn to secrecy about the problem. The little ones breakfasted in their pyjamas, oblivious to the tragedy that was unfolding.

After what seemed like forever we saw Sarah's estate car coming down the road. Clive filled buckets with warm water and the three of us set off back towards Queenie, armed with ropes and a tack box full of different implements. Queenie hadn't moved since we'd left her. Once again I put the halter on her, and Sarah began her examination.

'Yes, yer right, it's coming backwards,' she said, adding that it would be extremely unlikely that we'd get a live foal in this situation. 'But we don't know for sure. I'll do my best.'

The foal needed to be pushed forward so that the rear legs could be brought up from below and unfolded into a position in which it could be delivered. There was not a lot of room for manoeuvre, but working very skilfully and bringing one leg up at a time, Sarah eventually managed it. Tying a rope to each of the hind feet, she started to pull. A lot of strength was needed, and Clive took over pulling on one of the ropes. Queenie lay down at this point and I knelt beside her head, stroking her gently as she pushed in silence. Clive and Sarah pulled; I concentrated on her morale. 'C'mon, Queenie,' I said. 'C'mon.'

Moments later, the foal arrived. Pulling it quickly away from Queenie, the filly was laid out on the grass. I hoped beyond hope that there would be signs of life, but Sarah hunched over the newborn, then turned to me and shook her head. Clive stood, hands on hips, looking down on what could have been. Queenie exhaled and sighed. Tears welled up in my eyes.

'I'm so sorry,' Sarah said. 'The foal was already dead, been dead a little while.'

'Just one o' them things,' said Clive.

'What about Queenie?' I asked Sarah tearfully. Had the birth done irreparable damage to her?

'I can't see any reason why she won't recover from this,' Sarah replied. 'I'm going to give her a big injection of antibiotics to ward off any infection.'

Queenie stayed lying down while Sarah injected the penicillin. I looked across towards the house where the children were now all lined up on the top of the wall, some still in pyjamas, some in school uniform. I didn't wave, just looked back down at Queenie. Sarah was now gathering up her equipment, and washing her hands in the bucket of warm water.

'I just wish that I'd done summat sooner,' I said.

'You couldn't have known,' said Sarah. 'And the chances are that the outcome would've been the same, anyway.'

It didn't lessen the pain, and all the 'what ifs' still raced through my mind.

'What shall we do wi' t'foal?' I said, directing my question to stoical Clive. I could hardly bear to look at the body: it seemed so unfair, such a waste.

'We'll leave 'er t'foal for a while,' he said. 'That'd be t'best thing, then she'll understand.'

I was convinced that Queenie knew full well that her foal had died. Something in her demeanour, her quiet acceptance of our help, made me think that she knew before we intervened. But I was in absolute agreement that she should be able to say good-bye.

Clive and Sarah walked back to the farm while I took Queenie's halter off. Raising her head from the ground, she looked back to where the foal lay, her bottom lip quivering. But not once did she call for her foal.

'I'm so sorry,' I said to her gently, tears streaming down my face. Then I, too, walked away.

The older children were getting in the school taxi by the time I returned, puffy-eyed, to the farmyard. They knew what had happened and they were understandably upset. As I've said, children are extraordinarily resilient and we've never glossed over the truth or minced our words, so I gave them the details.

'It's a shame,' said Raven. 'Was it a bonny foal?'

'Aye, certainly was,' I said sadly, trying to keep my composure. Reuben and Miles were quiet.

'Do yer need me to dig a . . .' Reuben began, but the taxi door shut and they were off to school.

The day dragged. We'd missed breakfast, had no appetite at dinner-time, we seemed to have lost all momentum. The smaller children played in the sunshine in the yard, their happy laughter a direct contrast to the sad scene at the other side of the wall. I watched Queenie all day. Shortly after I took her a bucket of water, she lumbered to her feet and turned around to inspect her foal's lifeless body. She sniffed the corpse a couple of times, but never licked it. In the past I've watched yows desperately pawing at a dead lamb with a front hoof, willing it to get to its feet. But

Queenie seemed to understand that it was hopeless. By mid-afternoon she had moved away from the body and was standing in the relative shade of a mountain ash tree that stood in the corner of the field.

'We need to move t'foal,' I said. 'Queenie's not botherin' wi' it now and there's a few flies around.'

'I'll ga an' put it in't'bike trailer,' Clive said. He laid her to rest in an unmarked grave at the moor bottom.

Queenie staged a remarkable recovery over the next few days: she regained her composure and began grazing again. A week later and we decided that she should be allowed out of the Low Bobby Dale and back to the moor, as only Josie was there and we thought a bit of friendly company might do her good, revive her spirits.

I gave birth to my eighth child, Clementine, on my own in front of the hearth by the fire one June night, with Clive and the children sound asleep upstairs. The powers that be had dictated that, as usual, I should call for an ambulance as soon as I felt the birth was imminent. I don't have long, arduous labours, for which I am thankful, but it means that I don't get much warning.

Of my other seven children, only two were born in hospital. One was born prematurely at home, and the remaining four were all born en route to hospital. Realistically, the chances of me getting to hospital this time were slim, but I was forbidden from having a home birth because I was too far from medical help should something go wrong: the midwives would not be allowed to attend me during the birth, their insurance would not cover it, nor were they permitted to leave any basic birthing equipment with me in case it appeared they had given me the tacit go-ahead for a home delivery. If I went into labour at Ravenseat then I'd be on my own . . . literally.

I saw the midwife the previous Thursday, we talked about the proposed birth plan, and I was typically evasive. Never a big lover of plans, I reluctantly agreed that if the baby didn't turn within the next couple of weeks (it was lying transverse) I'd have to be admitted into hospital, because babies cannot come out sideways. I left the surgery on that gloriously hot morning feeling fat, full of baby, slightly perturbed about the prospect of going into hospital, but otherwise happy. There was little point worrying; things always sort themselves out. The beautiful weather brought with it a stream of visitors to Ravenseat. I signed copies of my first book and ran back and forth with cream teas, my bump serving as a platform to rest tea trays on.

On Saturday, midway through the afternoon, I began to feel slightly unwell, so Clive and Raven took over serving the teas while I went upstairs to lie down. I had some paperwork to attend to, so I didn't feel I was entirely skiving off.

'Are you OK?' asked Clive, poking his head round the corner of the bedroom door when he had a quiet moment. 'Are yer gonna 'ave t'babby?'

'Nooooo,' I said. 'I'm fine, just a bit knackered.'

I wasn't lying. I didn't think that I was going to have the baby, although unbeknown to Clive I did shove a few essential bits and pieces into a bag just in case: nappies, babygro, another stretchy dress, lipstick, bar of chocolate.

'Well, you just keep yerself quiet then, 'cos I got us a baby-sitter for tonight,' he said. 'I reckon that yer should book us a table at t'Black Bull.'

Babysitters are pretty thin on the ground around here, so this was not an opportunity that I was going to miss. It perked me up no end, the prospect of a night out with friends, and by the time Steven arrived to babysit I was feeling as fit as a flea. The downside of having Steven babysit was that in return we would have to dogsit Parsley, his canine sidekick, for a few weeks whilst he went to Ibiza to promote the Hard Times club brand.

Clive and I got ourselves tidied up. I squeezed myself into a long black dress and stood sideways, looking in the mirror. I was very round, there was no disputing that, but I thought that the bump looked different. I didn't analyse the reflection for too long: I'd spent the last few months being told that my bump was either high or low, that I was either enormous or neat, that I was clearly having a boy . . . or possibly a girl. I slipped a voluminous coat on, we said goodbye to the children, and headed off out.

I ate like a pig that night: starter, main course, dessert and anything that was left over. I hadn't been that hungry for weeks.

'Jeez, yer sure it's not just pies you've got in there?' said one of our less tactful friends.

We had a good evening and were soon heading back home: me in the driving seat, wedged behind the wheel, a tipsy Clive talking some rot about times past with me half listening. We slept well that night, and the next morning Steven departed, leaving Parsley behind. A typical Jack Russell, smooth-coated tan and white with pricked lugs, she had always been quite a territorial little dog, acting as Steven's car alarm. Nobody would dare to go near his motor if Parsley was in there. She'd furl back her lips, bare her teeth and growl and snap. What we hadn't realized was that old age had crept up on Parsley, and she was now partially deaf and not as agile as she had once been. It wasn't fair to put her in the kennels with the sheepdogs. She was fine just mooching around the farmyard during the day with our terriers Chalky and Pippen, but at night they slept outside in the barn, on alert for any intruding rodents, and we didn't reckon that the geriatric Parsley would care for sleeping under the stars. So she made her bed in our pickup. The front bit!

We were busy with guests staying in the shepherd's hut. Parsley soon learned from the other terriers that hanging around the picnic benches when cream teas were being served could be extremely rewarding.

We spent Monday morning moving yows and lambs about to different pastures, and trying to get some order among the flock. In the afternoon I nipped over to Hawes to pick up a prescription for iron pills, as I was slightly anaemic; then Clive took Miles and Reuben to Muker Reading Room for band practice. They were back by nine, and not long afterwards Clive and I went to bed.

I couldn't get to sleep, which was not unusual. My mind sometimes goes into overdrive at bedtime, thinking about the next day, working out what needs to be done, what I mustn't forget: the list is endless. The house finally fell silent and I lay there, folding my arms across my tummy. I wasn't uncomfortable, but sleep eluded me. I decided to get up and make myself a hot drink. I crept downstairs, made tea and sat down on the sofa to read a weekend newspaper. Usually this would send me to sleep, but not this time. The fire burned down to dying embers and needed reviving, but I didn't want to wake everyone with the crash of the coal scuttle being emptied onto it. I knelt on the rag rug in front of the hearth, leaned over the fender and picked small lumps of coal out of the scuttle with my hands, piling them carefully on the grate. I washed the soot off my hands, then on a whim decided that, as it had been such a beautiful day, I would go outside to see whether the sky was clear. I'm not a dedicated stargazer, but up here we have no light pollution and I often look at the heavens when it's dark. As I opened the door Pippen slipped into the kitchen; the wily old thing had heard me stirring – she'd been lying in wait to claim her place in front of the hearth.

It was a bonny night, very still, not a murmur of wind, and I was tempted to take a stroll down the farmyard. I thought better of it when I remembered that we had guests in the shepherd's hut who might find it unsettling to see a skimpily dressed person complete with wellies wandering about in the dark. I shut the door and went back to the fire, which was now crackling as the coal took light. I looked at the clock and sighed to myself. It was just after midnight, and I was going to be really tired in the morning if I didn't get some sleep. Back upstairs I sneaked into bed, dozed a little, but hadn't been there long when Annas tumbled out of her cot bed. Her shriek woke me, and after scooping her up and tucking her back into bed, I decided that I might as well go down to the sofa again and read more of the newspaper. Annoyingly I now seemed to have indigestion.

Only when I got back downstairs and went into the living room, sitting barefoot in the dim light, shadows being cast by the flickering flames of the fire, did I sense what was actually happening.

I'm in labour, I thought, feeling my tummy through my nightdress with my hands. *I'm going to have the baby tonight.*

I looked at the clock. It was a quarter to one. All was quiet in the house, nothing stirred. Pippen had moved from in front of the fire to under the settle, and was now fast asleep. It was time to make a decision. I could wake Clive, which would inevitably lead to the children waking up, ensuring a general panic as the ambulance was summoned and the hospital alerted, everyone rushing and hurrying. I knew I would spend my labour on the phone to ambulance control, where some poor operator would try to talk me through a normal delivery, with all the questions about timing my contractions (I don't have them), and instructions to lie down and wait for the paramedic. It would all be very predictable. I'd done it before.

Or there was an alternative. I could do what I had talked to the midwives about. I could do nothing, put my trust in nature and give birth alone, unaided. Freebirth, as is the modern expression.

'I were tupped at Ravenseat an' I'll lamb at Ravenseat' seemed to be the way to go. All I had to hope was that the baby had turned, and was coming right. I cast my mind back to Saturday, how I'd felt ill in the afternoon, how I'd thought that my bump had changed shape and how I'd wolfed down all that food after weeks of not being able to eat much. By my estimation, it all pointed to one thing: the baby HAD turned.

That was it – my mind was made up. It was time to make some preparations. I got the scatter cushions off the sofa and arranged them on the floor in front of the fire. As usual there was a pile of clean washing on the window seat waiting to be put away, and in and amongst it were some clean towels. I draped these over the cushions and then perched on the fender. A few minutes later I had an overwhelming urge to go to the toilet, but thought better of it.

Squatting over my makeshift mattress I pushed hard. Gravity perhaps helped, but that was all it took: one push and I felt the baby emerging. The moment of truth had arrived: I hoped and prayed that it was a head. My prayers were answered – in my hand I could feel the very top of the baby's head. I pushed once more, and this time I had both hands at the ready to catch the baby as I delivered her. I looked down to see my newborn's face looking right back at me, illuminated in the glow of the fire. I cradled the vernix-coated body with one hand whilst my other cupped the back of the tiny head, and then I brought the baby up to my chest. I rose, perched back on the fender and just held the snuffling baby tightly. Time seemed to stand still, a few precious private moments together before I had to think of the practicalities of the situation.

The baby was still attached to the cord and I hadn't yet delivered the placenta. I had once been warned by the midwife that there was a real danger of bleeding to death if I didn't get on with it, but it came easily. I put it onto a towel and then wrapped it up. It was only at this stage that I realized that I didn't know whether my baby was a boy or a girl, so, carefully moving the still-attached cord aside, I looked. A girl.

I needed to juggle things a bit now, as I didn't want to cut the cord. I've always believed in leaving the placenta attached for as long as possible. Grabbing another towel, I wrapped the placenta and the baby up together in one big pink bundle. I looked at my baby girl, her tiny fingers peeping over the hem of the towel, and stroked her cheek with the back of my finger. I was euphoric, but also suddenly incredibly cold. I was shivering, my teeth chattering. I couldn't put any more coal on the fire as I had my hands full and, besides, it was probably time that I introduced the new arrival to her father.

Up the stairs I went, carrying my precious bundle. It was less than an hour since I'd come down feeling restless. So much had happened, but in a strange way it also felt like nothing had happened: it felt so natural, normal, organic.

'Clive,' I whispered, reluctant to break the peace. He slept on.

'Clive,' I said a little louder. I didn't want to startle him. He shuffled in the bed.

'Clive,' I spoke at normal volume now. 'Wake up. I've 'ad t'baby.'

At this he sat bolt upright, squinting in the half-light. What he said is unrepeatable, but it ended with, 'Yer some woman, thee.'

We went downstairs, me smiling like a Cheshire cat, Clive quiet, still shell-shocked.

'I'll get t'kettle on,' he said, after banking up the fire. 'Then we need to ring the 'ospital an' tell 'em what's 'appened.'

I pulled the throw off the sofa around my shoulders while I got the baby latched on to my breast. Clive reappeared, carrying a cup of tea.

'What is 't?' he said.

'A girl . . . I think.' A moment of doubt crept into my head. 'I'm sure it's a girl.'

'Let me 'ave a look,' he said, unravelling the towel. I was right.

'Now, where's yer notes an' stuff? You need to ring the 'ospital.'

I wasn't keen and pleaded with him to wait until daylight, but my protests were ignored and he insisted. A midwife at the Friarage Hospital at Northallerton took my call. We went through the usual rigmarole. Who am I? Date of birth? Then the big one, what's your due date?

'Erm,' I tried to evade the question, because I knew what was going to happen. 'It'll be on yer computer, won't it?' I was being vague.

'Oh, 19th July. Dear me, that makes you five and a half weeks early. You can't come to us, we do not have a special care unit – you need to go to James Cook Hospital at Middlesbrough.'

My heart sank. I knew that they had to make sure that the baby was well, but it took nearly two hours to get to the Friarage at Northallerton, and that was far enough to go at the best of times. James Cook was even further away; you could add an extra half an hour for that journey.

'Do we 'ave to go now?' I said, flatly.

'Yes, I'm going to call an ambulance,' she said.

In the two hours it took for the ambulance to get to Ravenseat I left Clive holding the baby while I got showered, put the towels in the washing machine, cleaned up the trail of drops of blood between the bedroom and the living room, and dug out the baby car seat.

'Where yer gan?' Clive said, as the baby and I were escorted to the ambulance.

'James Cook, the hospital at Middlesbrough,' I shouted. 'Do yer know where that is?'

'Thomas Cook? Did yer say Thomas Cook?' he replied.

'Noooo, I'm not goin' on bloody holiday,' I shouted again, as the door slid shut.

It was after five in the morning and the sun was rising when we arrived at the hospital, and I was admitted onto the maternity ward. After being checked over we were declared fit, and it was time to think about how to get the sixty miles back home. I had planned for every eventuality and had cunningly brought the baby car seat with me in the ambulance, as I knew full well that I would not be allowed to leave the hospital without one.

Outside the maternity wing there was a bus stop and from there, every hour on the hour, a bus ran to the hospital at Northallerton. Perfect. We were on the eleven o'clock bus, which got us to Northallerton in half an hour. A short walk down to the bus stop outside the Tickle Toby Inn, and we were on another bus to Richmond. From there I should have been able to get back to Keld on the little white bus that ran through the Dales, but unfortunately I missed the connection. We sat in the cobbled square on the steps to the war memorial in the sunshine, and waited for Clive to come and get us.

It had been an eventful twelve hours: I'd missed a whole night's sleep and I was beginning to flag, but I was looking forward to showing the children the new baby. They had all slept through the events of the night, including the arrival of an ambulance in the early hours. Clive told them all the news at breakfast. Everyone was overjoyed apart from Sidney, who cried because he wanted a boy baby.

The midwife visited Ravenseat every day, weighing the baby – who now had a name, Clementine, or Clemmy for short. Unfortunately, due to her early appearance, Clemmy went yellow with jaundice and lost marginally more than the acceptable 10 per cent of her body weight. So it was back to the hospital at Middlesbrough, where she lay under a blue light for a couple of days.

It was just sheer luck that poor Parsley didn't get to spend a couple of days at the hospital too. Unbeknown to me, she had jumped into the pickup whilst I was going back and forth loading everything to take to the hospital, and had curled up behind the front seat. Clive turfed her out when he went to retrieve his coat just before I left.

When we returned from the hospital for the second time, Sidney was over his disappointment at Clemmy being a she-baby and was so thrilled with the new member of the family that he managed to trip over his own wellies and cut his chin quite badly.

'I can't believe it,' I wept as I rang the doctor's surgery. 'Please tell me that yer can sort 'im out, I can't face having to turn around and go back to Middlesbrough.'

Fortunately, armed with a plentiful supply of steristrips, our doctor patched Sidney up, and another trip to the hospital was narrowly averted.

And now the disclaimer.

I feel very privileged to have had what I consider to be a perfect birth. It may not be everybody's idea of the right way, but for Clemmy, me (and Clive) it was wonderful. I accept that I was lucky, and I don't go around promoting the concept of free-birthing. I simply believe that in my circumstances, it was the right option.

7

July

It was the weekend of the Tour de France in 2014. All the roads into Swaledale, where the route was going, were closed off. There was a sense of great anticipation, as thousands of people were expected to descend upon Swaledale to see the peloton as it passed through. Clive and I are not particularly cycling fans, but the fanfare and build-up to this once-in-a-lifetime experience meant that some of the excitement had rubbed off on us. We planned to take the quad bike across the moor towards Shunner Fell and then out onto the Stags Fell, avoiding the closed-off roads by travelling cross-country, and ending up in a prime location at the very top of the Buttertubs pass. From there we would be able to see the riders tackle one of the fastest, most dangerous stretches of the race.

The day before the big race I suggested to Clive that we should go for a walk after tea, a quiet moment for us as the children were playing together happily in the garden. We couldn't decide which direction to go, but finally chose the moor. Hand in hand we walked through the gate and followed the track up to the first hillock, and from there we could see Josie, standing alone at the moor bottom. It was a peculiar stance, almost as if she was standing to attention, square, upright, with her ears pricked. I frowned. There was something not quite right.

It was only when we ventured closer that we could see she was standing guard over Queenie, who had collapsed. Distraught, I ran to her side, screaming, 'Clive, we've gotta get 'er up!'

'I'm goin' for t'loader tractor,' he shouted. 'I'll bring some straps wi' me.' He hurried off back to the farmyard.

She'd gone down in a wettish spot and her white coat was now plastered with mud where she'd floundered.

'What the hell are yer doin' here? What are yer thinkin' of? You stupid, stupid 'orse.' I felt almost angry with her.

Clive reappeared on the horizon, smoke billowing from the exhaust of the tractor.

'We're gonna get you up and out of 'ere,' I told Queenie determinedly. Her ears flickered, but in her eyes I saw consummate tiredness.

Clive pulled up close, stopped the engine and threw two ratchet straps to the ground. 'We need these under 'er belly an' out t'other side,' he said.

Dropping to my knees, I threaded the first strap under her brisket and out the other side whilst Clive pushed the other strap further under her abdomen.

'Reet, I'm gonna lower t'front end loader down, an' you're gonna loop these straps around the spikes,' he said. I nodded. 'Yer gonna 'ave to be careful that she don't fall on yer, but yer gonna 'ave to try an' encourage 'er to support 'er own weight, too.'

We tried three times but we just couldn't get Queenie up. It started to rain, and soon both Clive and I were caked in mud.

'Mand, it's naw good, she's nivver gonna stand.'

'Please, please . . . just one more go,' I begged, although in my heart of hearts I knew we were fighting a losing battle.

'Yance more,' he shouted. 'Yance more an' that's it.'

We failed. She was just too weak to try any more, and as the rain now came down harder we were both sitting there sopping, wet to the skin. Queenie's mane that I'd patiently combed and conditioned was now knotted and tangled and stuck to her neck with mud. We could see it was futile.

'We need the vet out,' said Clive softly.

'I'm gonna ga an' get 'er rug,' I said, speaking quietly and smiling down at her. I was biting my lip to stop myself from crying. Josie was standing close, watching.

I went back to Queenie's stable while Clive called the vet. When he came out of the house he was frowning. The vet was on a callout at the other side of Appleby, was about to perform a caesarean on a cow, and the road into Swaledale was closed because of the Tour de France. She suspected that the sudden collapse might have been due to a blood clot after the traumatic labour and stillbirth, but whatever the cause, it was clearly the end of the road for poor Queenie.

'There's only one thing you can do,' she said.

Taking the children into the house, I put on music and turned it up, not wanting them to hear the shot ring out. Clive went to the gun cabinet while I picked up a halter and walked slowly back to Queenie. So many thoughts ran through my head: had I explored every avenue? Was there anything else to be done? Worst of all, was it my fault?

I covered Queenie with the rug, and straightened out her tail so it lay neatly across her tucked-up legs. I looked her in the eye. The same watery blue eyes that had been filled with sadness, vulnerability and fear when she first came to Ravenseat now looked quietly content. I ran my fingers through her forelock tenderly, and then gently tucked it behind her ear. Then I walked away, never looking back. I slipped the halter onto Josie, who quietly walked behind me back to the farmyard. Halfway down the track I met a pale-faced Clive, his gun over his shoulder. We passed each other, but nothing was said; only the clip-clopping sound of Josie's hooves cut through the silence.

Queenie was buried where she fell, which was exactly the place that, unbeknown to me at the time, Clive had buried her foal. It made we wonder whether Queenie knew that her foal was there and had chosen this to be the place where she died, so they could be together again. That might just be sentimentality and imagination on my part. But as a mark of the high esteem we held her in, Reuben made a headstone for Queenie, with her name on it.

I was bereft after her death, on the verge of tears much of the time. I talked to my friend Rachel about what had happened and how it was perhaps a modern concept to grieve so much for a horse. Maybe nowadays horses are more cosseted than they were in the days when they worked on farms, when folk were maybe less sentimental about them. She said that there was possibly some truth in this, but she had something she felt would interest me. Sure enough, a few days later, the postman brought me an envelope from her.

Inside, on a scrap of paper, was a poem about a funeral held at Birkdale, which is just at the end of our road. It is dated 1st October 1910, and is called 'The Interment of Daisy'.

Old horse, old horse, how came'st thou here?
Thou has roamed the fells for many a year
Until by one unwary slip
Into the Swale thou got a dip.
And being old and not quite slim
Thou accidentally broke a limb.
Thus it fell to thy sorry lot

To end thy days by being shot.
Thou'st ne'er had blows or sore abuse
Or been salted down for sailors' use.
To eat thy flesh and pick thy bones
And send the rest to Davy Jones.
So ends at last this sorry tale
Thou'll rest in peace beside the Swale
And when the last great trump shall sound
Thou wilt spring to earth at one great bound
To show that thou art fit and smart
In a future life to take a part.
A lady graced the funeral scene
And kept her boys at peace
Who gambolled on the turf and green
She bid their antics cease.
The day was fair as fair could be
As we walked sadly o'er the lea.
Gentles and simples, saints and sinners
All trooped off to a jolly good dinner.
—Wrong Fellow

I loved the notion of this – another horse, over a century earlier, being held in such high regard that she warranted a proper funeral.

There are two major jobs in the summer months of July and August, clipping and haytiming, and which we do first depends on the weather. Mowing the meadows can't start before mid-July, as the environmental schemes we take part in are designed to give the wild flowers time to set their seeds.

We usually start by clipping the hoggs, who are now about fifteen months old. As this is their first clip, they are carrying the greatest weight of fleece, with plenty of new growth of wool, which we call 'the rise'. The yows have put all their energy into milk production to feed their lambs, and will not have grown any new wool yet, so they are clipped later.

If it looks like we're in for a good spell of weather, we'll do the hay, but we really need to get the clipping done by the end of July. I love clipping, but Clive is not as enthusiastic: he says it's a young person's game, which is why I enjoy it. I'm not sure if I'm getting faster at clipping or whether he's slowing down, but either way, we get the job done. Clipping the hoggs is a tough way to start – not just because they are very woolly, but, being new to it, they struggle and wriggle. A well-behaved hogg can have its wool removed in a couple of minutes, but a naughty one can take twice as long. The number of times that I've heard Clive muttering, sweat dripping from his brow, 'I coulda done yer, yer daft beggar, will yer sit thi'sel' still . . .'

It takes strength and skill to keep a hogg in place, and their sharp horns can do your legs some damage in the process. I tell myself that clipping equates to a pretty intense workout and that by the time we've done near on a thousand sheep I should have a fit, toned body. Not that I can show it off: however hot it is, minidresses are out of the question, as shins, calves and thighs will be black and purple with bruises. Clipping a pen of a hundred hoggs is a daunting prospect, not least because of Clive's dodgy counting.

'Just a hundred for today, Mand,' he says.

'That's seventy done now,' I say, looking at the still suspiciously full pen of woolled sheep waiting their turn. 'Are you sure there's just a hundred?'

'Thereabouts,' he says.

I suppose it is better for morale to underestimate.

'They're only to clip yance,' he says.

Of course, before we clip we have to round them up. This means a very early start. Gathering on the moor requires dog power, and if the weather is hot then the dogs soon tire. Avoiding the heat of the day works better: the sheep sense if the dogs are flagging, and take advantage by breaking for freedom. The sheep know their own heafs, and all the dodges. It is very frustrating to have a good gather and be homeward bound, only for some to break away and head back to the hills while we are powerless to stop them because the dogs are totally exhausted.

We have a system for gathering from the common, where others run their sheep. We don't cast our net too wide, in order not to bring in too many yows belonging to our neighbours. We work closely with the other commoners, but invariably end up with a few strays. They, too, will end up with some of our sheep. At certain times of the year there is much toing and froing to the outlying sheep pens dotted around the moors, where the strays are left until their rightful owners come for them. Everyone will gather on the moors at roughly the same time because we all have the same jobs to do: clipping, speaning, tupping and lambing are the four big dates in the sheep diary.

We try to clip the sheep as soon as possible after gathering them, as they don't settle so well in the fields around the house, and they soon eat all of the grass available to them. A hungry sheep is an unhappy sheep, and clipping hungry sheep makes for an unhappy clipper. A full sheep is round and fat and you can easily run the clippers over her curves. A thin sheep is more angular, with wrinkles and bony bits making the job far more difficult. Putting the sheep overnight in the barn guarantees they will be dry the next day, but the downside is that they will also be likely to have straw stuck in their fleeces and muck clagged up in their hooves. The less contamination of the wool, the better.

Once upon a time, wool was an incredibly valuable commodity and the wool cheque would pay the tenanted farm's rent, so great care was taken over the wrapping and packing of the fleeces. Not any more: the poorer-quality, coarser wool from the hill breeds is no longer in demand, and in return for the 2,000 kilos of wool that we send to the wool depot we can expect a payment in the region of £400 – in a good year. One year, our wool cheque was for the princely sum of £65. Some farmers have resorted to burning their wool clip; the natural oils in it mean that it will burn when freshly shorn. Somehow that doesn't sit right with us: it seems so wasteful.

Weather conditions throughout the year influence the amount and quality of the wool produced. A bad spring, when the yows haven't been as fit as they normally are, means that they will cast their fleeces when they begin to thrive again. The same goes for a sheep that has been sickening: when she begins to recover, she will throw her wool off. One sheep in a field losing its fleece makes a terrible mess. In the old days, this wool was never wasted, being collected and used by the farmers' wives to spin into yarn. Even within living memory, wool was prized, and Clive remembers being sent to clip a dead yow.

Clipping time at Ravenseat can be finished in a week if the weather is good, and it really is a family affair. The smaller children bounce in the wool sheets to pack the wool down tightly, while the bigger children help us catch the sheep. A catcher makes the whole job of clipping a lot easier, as they bring the next woolly customer right to you. This means you don't have to turn off the clipping machine, don't have to hang up the handpiece and don't even have to straighten your back. A supply of sweeties is a necessity in order to curry favour with the catchers, for they select your next sheep, with the bare-necked and bare-bellied ones going down a treat, while any with mucky tails or fleeces full of grit and soil do not.

In a hot summer the sheep will become itchy in the heat and take shelter from the sun under the overhangs of the peat haggs, where they have a good old scratch. The soil and grit finds its way deep into the fleece, and makes the combs and cutters very blunt very quickly. A rainy summer makes for cleaner fleeces, as the rain rinses the muck away. My latest baby usually sits in a pram watching the clipping, or is lulled to sleep by the constant hum of the machines.

Raven is learning to clip. It's not all about strength but technique too: keeping the sheep moving so that it doesn't feel the urge to struggle, and being able to hold the sheep with your legs. Clive and I both clip the old-fashioned way, from the back of the head and down the shoulder. The modern way of clipping, from the bottom leg upwards, is quicker, but it's impossible for me and Clive to change now. Our dilemma is whether to teach Raven to clip our way, or whether she should go on a shearing course and learn the newer method.

While we are clipping the yows, the lambs are held in a separate pen and then reunited with their mothers afterwards. When they've been clipped we give them time to mother them up with their lambs again, as without their fleeces the yows look and smell quite different. If we turned them straight back out to the moor then we would inevitably end up with some lambs being unable to find their mothers and being speaned at three or four months old. They can survive without their mothers' milk, but will not do as well living only on the rough moorland grass. It's far better to take great pains with them and make sure that all the yows have found their lambs.

One thing that keeps me going during this physically tough job is the thought of a swim in my own personal hydrotherapy spa. The dirt and sweat, and the aches and pains from the toils of the day, will be gently washed and massaged away by taking a dip in the dub, the river pool at the back of the farmhouse. Fully clothed, I swim a few lengths and then plunge under the waterfall. I emerge feeling cold, fresh and very much alive, and then hang my jeans and vest on the washing line ready for another day.

We keep a tally of how many sheep we have clipped, and usually we find that we are a few short. Invariably they will turn in late, perhaps on the next gather for speaning in September, or occasionally even later at tupping time. Very occasionally one misses getting clipped altogether and ends up with two years' growth of wool, but this is bad for the sheep as they get weighed down with snow during the winter, then blown over and rigged (stuck on their backs upside down) because of it. If a woolled yow turns up in September we go back to the old, traditional way, clipping with the hand shears on a clipping stool. Using the hand shears means that we can leave her with more wool for warmth, whereas with the electric clippers it's either all or nothing.

Life on the farm brings a heightened awareness of the passing of the seasons. They, and the jobs that go with them, come and go with worrying frequency. There is no ideal time to have a baby, and I have long since given up on family planning around the farming calendar. In theory I don't want to be in the late stages of pregnancy in winter, when there's a chance of being snowed in, and I don't want to be in the early stages during clipping and haytiming because of the strenuous physical exertion needed. In a perfect world, I don't want to be at any stage of pregnancy during lambing time . . .

As with Annas, Clementine was born early, in a hurry to get here, and her early arrival meant that my postnatal exercises were once again done in the sheep shed, clipping. For some inexplicable reason we had also agreed to appear in a TV programme with Ben Fogle. Clive had liaised with the film crew on the telephone and invited them to come and see us before the filming. I overheard Clive's conversation with the producer on the phone one day:

'I think it would be a good idea for me to come and meet you and your family before we begin to film,' said Tina, the producer. 'Then I can also get a feel for the place.'

'A varry good idea,' said Clive. 'Yer could come up fra' London on t'train, stay at t'hotel in Hawes, then go yam the next day.'

'Well, I think I'll just set off early in the car and go back at teatime,' said Tina.

'Whhhhhhhat?' said Clive. 'Drive? All the way from London to Ravenseat? And back?'

'Yes,' said Tina.

'Well that's varry brave o' yer, Tina,' he gushed. 'To drive all that way. That's 'ellish.'

I turned round and scowled at him. When he put the phone down, I treated him to a piece of my mind.

'I can't believe yer think that driving a few 'undred miles in a car is that bloody brave when yer own wife gave birth on her own in t'living room only last week.'

One of the ideas for the programme was that Clive and I would be clipping sheep in the pens and Ben would come and help and have a go at clipping, catching and turning up the sheep. In theory it all sounded OK, but I had a few worries, chief amongst them: 'I can't get mi shearing jeans on,' I said miserably. 'I can get mi legs in 'em, but not mi arse.'

Clive had a few worries too. 'I need to 'ave mi best sheep in t'pen,' he said. 'I can't 'ave 'im clip owt 'orrible on TV.'

'There's gonna be a terribly 'andsome TV presenter clipping, and me, in my buxom state spilling out of my vest wi' my jeans fastened shut wi' string, so I don't think there's gonna be so many folks lookin' at thi bloody sheep.'

Then I reconsidered, and we put aside forty of our very best hoggs.

The string held, thankfully, and the sheep were clipped on camera. I dare say it wasn't a pretty sight, but that's reality for you.

The baler twine saved the day yet again; it's not called the 'farmer's friend' for nothing. Rummage in the pocket of any shepherd and you'll find, along with a knife, a length of baler twine. Its uses are innumerable. It can fasten a gate, tie up a sheep, hold up a pair of jeans or act as a belt when my baby bump gets too big for my coat to fasten. During a recent conversation about how invaluable it is, I was told about a farmer who used it instead of luggage straps, always fastening it around his battered suitcase when he went on holiday. It was easily spottable on the carousel at the airport, and nobody ever took his bag by mistake.

We don't go on holiday, but we do manage to have occasional days out, and after our family appeared on *The Dales* television programmes a few years ago we were invited to attend a special lunch at the Great Yorkshire Show. The showground, on the outskirts of Harrogate, is spread over 250 acres and is the biggest agricultural show in England, with over 30,000 visitors over the three days it is on. The Yorkshire Tourist Board invited us and arranged everything, from car parking tickets to members' badges, and even a whole table reserved for us in a beautifully decorated marquee.

We only had five (or was it six?) children at the time, but that's enough to shepherd through the show to the marquee. There were too many distractions.

'Ooooh, look, a pen of donkeys,' one of them would say, and the rest would trot off to have a look. Then: 'Sweeties, lots of sweeties,' and they'd be off again, just as we'd rounded them all up.

Every so often we did a head count. Yes, all present and correct. The only one I couldn't actually misplace was the one that I was carrying on my front, hidden beneath my jacket. The lunch was very good, but not relaxing. There was much arranging of napkins in case of spillage and we missed Chalky and Pippen, who enthusiastically clean up the floor at home when the children have eaten.

The meal ticket was not completely free. I had to give an interview to the local radio station, mingle with some of the great and good, and have my picture taken. After the last spoonful of dessert had been downed and the last glass of cordial supped, it was time for me to leave the family for a while.

'I've got to ga an' do mi stuff now, you're in charge for an hour or so,' I said to Clive.

He decided that he wasn't going anywhere: if he just stayed put in the marquee then the ladies from the tourist board would look after him, plying him with cups of tea and entertaining the children into the bargain.

I was soon back, and decided it was time for some retail therapy. Clive wasn't thrilled at the prospect of trailing behind me down the aisles of the huge show, but reluctantly took charge of my wonky three-wheeled running pushchair – the sort used by exercise-fanatic parents. Among other things, I used it when I went mole-catching around the fields at home. The traps went in the bottom underneath the seat, meaning the baby inside was not showered with soil, as they would be if I carried them in the front papoose. It wasn't ideal for sightseeing at the Yorkshire Show, having splashes of mud on the canvas and splodges of chicken poo on the sun visor from the hens roosting above it in the barn, but it was marginally better than having to carry a tired toddler.

As we were about to set off we did another head count. Only four. One missing. It was Edith, who was four at the time. I didn't panic, assuming that she'd be playing in or around the marquee and the ladies from the tourist board would know where she was. They didn't. Everyone had seen her, but nobody knew where she had gone. I looked at the crowds of visitors to the show milling past, and groaned. Where should we even start searching?

'You stay 'ere wi't' children, afore we end up wi' everyone lost,' said Clive, setting off into the melee. He was back within minutes.

'We'll nivver find 'er amongst all these folks,' he said. I could see he was starting to panic, but trying not to show it for fear of worrying the other children.

At that moment an announcement was made over the tannoy, reporting that 'Edith of t'*Dales* TV programme' was missing, and if anybody saw her could they please escort her to the lost children's tent.

I was obviously pleased that everyone would be on the look-out for her, but also mortified that my fecklessness was now being broadcast to the masses.

Soon afterwards the walkie-talkie of one of the staff crackled into life, reporting that Edith had been found. We were all loaded into a show-ground buggy, usually reserved for royalty and other VIPs, and driven to the lost children tent. I was so relieved, and imagined her running headlong into my arms, so happy to be reunited with us. Instead, she was smiling at everyone, and couldn't understand what all the fuss was about.

After that, we decided to call it a day and go home while we still had them all in tow.

A few days later we were talking to one of our farming neighbours when he mentioned being at the Yorkshire Show.

'We nivver saw thi,' Clive said.

'Naw, but I saw your Edith,' he said. 'I spotted 'er in t'handicraft tent, she were lookin' around t'stalls. I took 'er to t'lost property tent.'

'Was she crying? Did she look unhappy?' I said.

'Nay, she were as 'appy as Larry,' he said. 'In 'er own little world.'

We thanked him profusely.

I've always been proud that the children are independent and don't hang from my apron strings, enjoying the freedom that life at Ravenseat gives them. But there's a time and a place for being independent, and the Yorkshire Show isn't it.

This incident reminded me of getting lost as a child, only in my case it was in a department store in Leeds. I'd gone up the escalator with my mother, then gone up another escalator and another and, lo and behold, found myself alone on the top floor. I can't remember if I cried but I remember a store assistant ushering me into a small room with a mechanical typewriter, giving me a tube of fruit pastilles and telling me that I could play with the typewriter until they found my mother. A little time passed, and I was thoroughly enjoying myself busily plinking away on the typewriter when the door opened and the store assistant came in with my mother.

'Is this your mum?' asked the assistant.

I furtively typed away and, glancing up, said, 'No, it isn't.'

I was enjoying myself too much to think about the long-term implications of denying my parentage. The matter was soon cleared up and I dare say that my mother would have had some-thing to say to me, but I don't remember any of that bit.

Nowadays I avoid department stores and prefer to shop locally, ordering some things online. But during the school summer holidays, there comes a day when I must load up the children and head for town in search of the dreaded school shoes.

This is bad on so many levels. The children live in wellies, as we do. They are comfortable, easy to get on and off, and can, I've found, be worn with any outfit. Once upon a time wellies came in only green or black, and were worn only with water-proofs, but nowadays it's quite acceptable and, dare I say it, chic to wear them with shorts and maxi dresses. Who'd have thought that we would find ourselves at the forefront of fashion? But for us, wellies are a necessity: even the footwell of the Land Rover is no place for sandals or trainers. Just getting from the front door of the farmhouse to the vehicle can mean negotiating a myriad of obstacles, including muck, mud and puddles.

In our family wellies are worn on the right feet sometimes, wrong feet more often than not, and are frequently mismatched. People often point out to me that my children are wearing odd wellies; but if one welly has a hole in, you needn't throw the pair away. There will usually be another odd one in the right size to team it with. All the children have gone through a stage of refus-ing to take their wellies off before bed. At the moment it's Annas's wellies that I prise off from under the duvet when she dozes off.

So we set off to town, in wellies, the older children gawping at shop windows, the younger ones walking into other shoppers, as they have not grasped the etiquette of pavements. They don't understand that you cannot stop dead, lie down and peer down a drain. You can't smile and say hello to everyone coming the opposite way. Sometimes I think that I should bring a sheepdog, Kate or Bill, to shepherd them in the right direction and steer them into the first shoe shop we see. On our last shoe-buying expedition we went into a sports shop to buy trainers for the boys' PE classes. We'd hardly got into the shop when Sidney began to shriek, rooted to the spot. He was looking down, fixated on the highly polished black floor tiles, as he could see his own reflection below him.

'C'mon, Sidney,' I said. Other shoppers were now staring.

The screams got louder, then he got down on his hands and knees, but still the little red-haired, freckled doppelgänger was staring back. I had to pick him up and carry him across to the seats where customers tried on trainers, and leave him there until we'd found what we wanted. He lay flat on the seats, occasionally peeping over the edge into the shiny black abyss below.

When it gets to actually trying on shoes, I hope and pray that the socks inside the wellies are half decent. We get through numerous pairs of socks and, in the same way that we don't throw away a pair of wellies if one is holed, I don't get rid of two socks if one is still wearable. Finding a matching pair is nigh on impossible, and the children often wear any combination of long ones, short ones, thermal socks, trainer socks in a range of different colours.

I watched with bated breath when Miles prised his wellies off for the shoe-shop assistant, who had found a pair of new shoes in the right size. There are usually socks, that's for sure; with luck, there will be two. On one occasion I pulled out a scrunched-up sock jammed in the front toe of his welly that he didn't even know was there. Stuck to the socks there will probably be straw, hay seeds and soil, which ends up on the shop floor.

When eventually we found footwear for all of them I haggled, to see if I could get a discount for bulk buying.

'What if they're not the right size?' the assistant said, when I asked for the same style in a size one, three and five. 'Do you want to try them all on?'

'Nay, they'll fit someone at some point, that's for sure.'

The summer holidays are a glorious time: it can't be true but we seem to have week after week of sunshine, with evenings spent down by the river, the children paddling and the dogs patrolling the riverbank. Water holds no fascination for them. It is there to be paddled in sometimes, but for the most part it is an annoyance that washes their discarded scooters downstream, or prevents the school taxi from getting across the ford.

Just below the farmhouse, on the banks of the meandering river, there is a sandbank. Coarse sand, eroded from the sandstone boulders and rocks further upstream, is deposited there on the inner bend of the river. Many happy hours are spent here at what we call the beach, playing with spades, buckets and toy tractors. It is just outside shouting range, so this is the first place I look if anyone is missing at mealtimes.

Recently the children have widened their territory, roaming as far as the abandoned quarry or even to the High Force waterfall, returning with tales of yellow-bottomed frogs and complaining about Chalky eating their picnics while they were making woven sailing ships from the seaves. Wherever they go, there's a terrier following, always hopeful of a discarded sausage roll or unguarded KitKat. According to the experts, chocolate can kill dogs, but Chalky hasn't suffered any ill effects so far.

The problem with these excursions into the wild places of Ravenseat is that items of clothing are often abandoned, then not found for days, weeks, months – sometimes never. I imagine archaeologists of the future having fun up here, what with Reuben's mechanical creations buried in various nettle beds, and discarded hats, sweaters and anoraks pickled in bogs.

It is great to see the children setting out on an adventure, their pockets and satchels filled with food, sometimes hand in hand, other times mounted on ponies, with the dogs at their side. If either Clive or I go with them then the trip usually turns into a job, as we spot things that need doing. Perhaps it is a gap in the wall or a water rail washed out, or a tup limping, maybe a yow without her lamb. Then we end up returning to the house for a sheepdog to round up the tup or the lamb. It's no good taking the sheepdogs with you for the walk, as working sheepdogs are equally bad at switching off and relaxing; their minds are constantly on sheep. One minute Kate or Bill will be heeling you, the next moment you'll see sheep running on the far horizon and realize that the dog is no longer at your side.

Even our retired sheepdogs have difficulty switching off. It would be nice to let them have a free run around the farm, wandering as they please, but unfortunately this is rarely possible. For a sheepdog, rounding up sheep is what life is all about, and if there are no sheep available then they find the next best thing – which may be chickens, cows, horses or children. It takes a tolerant horse to put up with a dog nipping at its heels, or even swinging from its tail, as Fan does. Rounding up chickens usually ends with the dog getting a chicken dinner – not good for the morale of the rest of the hens. Violet once had her legs taken right out from under her when Kate found herself with nothing to chase and decided that Violet, who was skipping across the field, needed to be stopped in her tracks. Kate put in her best outrun and came in just a bit too close, fully expecting Violet to turn and run in the opposite direction, which is what a sheep would have done. She collided with Violet's legs, and Violet went down in a heap on the grass.

Sometimes we have walkers pass through with a dog on the end of a leash that looks to have the potential to make a good sheepdog. It will be eyeing up the sheep, watching their movements intently. I can't help thinking it must be difficult to keep a dog like this amused. I make no claims to being a champion dog runner, but I know from experience that a headcase of a dog, one that wants to chase aeroplanes, leaves and bike wheels, will make an excellent sheepdog, because you can channel that instinct. A dog that has no innate instinct to chase anything will never be a sheepdog.

The relationship between shepherd and dog is demonstrated both in working at great distance out on the open moors, and during closer in-bye (field) work. Sheepdog trialling is about displaying a range of those skills within the confines of a field: the gather, lift, fetch, drive, shed and pen. These are all skills used on a daily basis by a shepherd.

Just as there are legendary tups in the sheep world, there are legendary dogs in the dog running world. Among the famous names are Wiston's Cap, Dryden Joe and Hutton's Nip, the last a great favourite of Clive's because he was once lucky enough to see him in action. As a teenager Clive went along to sheepdog trials with his mentor Ebby, who was one of the top sheepdog handlers at that time. Clive remembers being in a large marquee after an international dog trialling competition when George Hutton was celebrating a win with a few beers. The tent was full of folk, and while George leaned against the bar talking he would without warning take his cap from his head and fling it into the crowd. Nip would set off into the sea of people, weaving in and out between their legs until he found the cap, returning with it in his teeth to George for him to throw again. It

was not done to impress: this was his way of keeping Nip enter-
tained while he got a few drinks in. Every so often George
would growl: 'Worry tha' bugger,' at which Nip would bite and
shake the cap.

George and Nip had the special relationship that comes from
being in each other's company and developing an understanding
of each other. A good shepherd can 'make' a dog, but a good dog
can also 'make' a shepherd. At sheepdog trialling events there
will be a variety of competitors: farmers, shepherds and other
enthusiasts who keep a few sheep, solely for the purpose of training a dog. Our friend Alec has his own small flock of sheep but
also likes to come and work his dogs at Ravenseat, which makes
him a very good friend to have. After spending his weekends
travelling to various dog trials held in fields on farms all over the
North of England, he tells us all the news on the dog front, and
whose dogs are on form. (It's usually his.) Sitting in the kitchen
at Ravenseat supping tea, his tales are so vivid it sometimes feels
like we've been there with him.

'I sent 'er left-handed, she mebbe got in a bit too close,' he'll
say. 'An' when we were comin' back down t'field yan o' em brock
away, then I sent mi dog reet . . .'

Sometimes I drift off and don't give his story my full attention. But there was one tale that did make me prick up my ears.
He was talking about a trial he'd been at where one of the dog
handlers, a chap of considerable age, had died whilst running his
dog.

'What 'appened?' I said, a little intrigued at this tale of woe.

The competition was in full swing, each handler waiting their
turn to run their dog on the sheep. As usual, the judge watched
from a distance and awarded marks based on the dog's perform-
ance.

The old chap's turn came and he walked out to the post in the field, sent his dog left-handed on the outrun to the back of the sheep, brought them back towards him, then round behind him, a tight turn. Then the dog executed a perfect cross drive, taking the sheep from one side of the field to the other. Guiding the sheep through the hurdles, he didn't put a foot wrong. Finally, all that remained was to put the sheep in the pen. Things were going extremely well, it was poetry in motion, a sure winning run. Man and dog were working together in perfect harmony: the sheep were moving towards the pen, the dog guiding them in slowly and under perfect control. All that remained was for the handler to close the gate behind them.

Suddenly, without warning, he dropped to the ground. At first the spectators couldn't understand what had happened; it took a moment or two for them to grasp that there was something wrong. Rushing across, they found the poor man had died.

'Bloody 'ell, Alec,' I said. 'That's awful.'

'Aye, 'e'd 'ad a bloody good run,' said Alec. ''Appen, best run o' t'day.'

'Well, at least 'is poor widow would get 'is winning trophy posthumously,' I said.

Alec dunked his biscuit in his tea and, looking up with a deadpan expression on his face, said, 'Nah, 'e'd nivver getten gate shut, he were disqualified.'

No room for sentimentality.

8

August

Traditionally the school summer holidays were timed to allow children to help their families bring in the harvest, and that's how it still works at Ravenseat. We grow grass for hay in about a dozen meadows, covering around a hundred acres, and we hope to make approximately 5,000 small bales. The meadows have names that have been passed down for centuries, but which you won't find on modern maps: Peggy Breas, Far Ings, Beck Stack, Black Howe, Hill Top, West End and the Hogg Hills. Some are named after their original owners, others are more to do with their locations. Placed along the drystone walls, which themselves tell of times long ago, are narrow stiles that allowed farmers on foot to get to their outlying barns, perhaps laden with a backcan to carry the milk from a cow tied up in a boose (stall). The smout-holes, square gaps in the bottom of the walls, were built to allow sheep to move freely between fields, but not cattle. Nowadays they are usually kept closed, with flagstones acting as doors.

In one of the meadows there is a fifteen-foot-square stone pen that can only be accessed through smout-holes, and we think this must have been made for sheep to shelter in when there's a blizzard, as there's no gate or stile for a person to enter by.

Many of the remoter, steeper fields that we now class as pastures, and use only for the grazing of animals, have barns still standing. They would once have been mown for hay with a scythe, but are not accessible or safe to negotiate with a tractor and mower. We still use some of the barns, keeping cows in Miles's cow'as and tup hoggs in the Beck Stack. These barns have water available, and are reasonably easy to muck out. Centuries have gone by, but the view remains largely unchanged and the aim is still the same: to harvest as much of the summer grass as possible to feed our animals with through the long winter.

We mow with a small rear-mounted mower on the back of a tractor. None of our fields can be cut completely because of the steep banks and gutters, but this grass is not wasted, as it is grazed by the newly speaned lambs in September. The children sometimes claim some of these uncuttable areas as their own hayfields, using a pair of garden shears, then turning the grass by hand to dry it and finally tying it into small sheaves with baler twine and then storing it in the loft of the woodshed. These tiny bales can be used as rabbit feed during the winter.

We get plenty of offers at hay time from would-be tractor drivers who fancy the idea of cruising around the fields, cocooned in an air-conditioned cab, with a front-end loader equipped with a flat eight to pick up the neatly arranged bales, never leaving the well-sprung seat to physically lift a bale. They imagine the bales stacked high on a large trailer and transported smoothly back to the farmyard. The reality is gripping the steering wheel with sweaty hands while trying to negotiate an impossible gradient, checking whether you are on wet ground by looking through the rust holes in the footwell of the cab at the earth below. Inevitably there will be an annoying buzzing insect in the cab, and on one occasion I had a small nest of chirruping chicks in the space that a radio had formerly occupied. That's if you are lucky enough to have a cab in the first place: our vintage tractors don't have them.

On a hot day you'll get sunburn, and on other days windburn, and after a typically long haytiming day there are other potential afflictions, like a crick in the neck from constant checking behind in case you have shed your cargo of bales coming up a hill, or even to check that your trailer is still attached. Then there is 'tractor-seat arse', which won't be found in any medical books but is well known to anyone who has spent a day bouncing up and down on a metal seat with no suspension. The symptoms include walking like John Wayne, and an outbreak of pimples on your backside. And if the weather is hot I have to choose between keeping covered up and overheating, or wearing a short dress and bare arms and being attacked by klegs (horse-flies).

Luckily, because we are hill farmers, grass is our only crop, so we spend limited time in a driving seat. This means that we don't invest in new, high-tech machinery. 'Make do and mend' are the Ravenseat watchwords, and we choose simpler, old-fashioned machines that are repairable with a crowbar and a hammer over anything electrical or digital. When you have broken down at 3.30 p.m. on a Sunday afternoon with rain imminent and a field of hay to bale, you don't want to be ringing a helpline that tells you to reboot the operating system, or that a vital part will have to be ordered from Hong Kong.

Once the hay is cut, the next job is time-consuming and mind-numbingly boring. 'Strawing out' to dry the grass is done with a haybob attached to the back of the tractor, a simple contraption that tosses the grass into the air and spreads it out. How often we need to do this depends on if the weather gods are looking favourably upon us. Clear skies, hot sun and a gentle summer breeze mean a couple of circuits will crisp the hay, but a 'slow' day means you can listen to the entire Queen back catalogue on your headphones and it still won't be dry.

'Wanna ga for a drink?'

'Wanna ga to t'chippy?'

The answer is always the same. 'Can't, I've got some 'ay down.'

It means there will be no relaxation until the hay has been baled and mewed into the barn. All the time it lies in the field, we fret about the weather. A few years ago it all hinged upon the Sunday dinner-time weather forecast on *Countryfile*. That gave a guide to whether we were going to bring in good hay, or get rained on and end up with big round bales of silage. Nowadays the weather forecasts come from many sources: the phone, the computer and there's even a TV channel totally dedicated to it. I think I preferred it before, when it was decisively and consistently wrong . . .

Now it's: 'Mand, just watched t'forecast on t'BBC an' it's gonna rain this aft,' says a perturbed Clive, looking at the sky.

'Naw, Met Office says it's gonna be dry,' I reply, looking at the iPad.

Then Alec will interject: 'It's gonna rain this mornin', boy,' he says. ''Eard it on t'wireless.'

There's luck involved in making the right decision.

Clive has the final say on when the hay is dry enough for the haybob levers to be altered from 'straw out' to 'row up'. Then it's back and forth making neat little rows ready for the baler to pick up. This is usually when the problems start. Once upon a time, back in the sixties and seventies, everybody made small bales that were easy to pick up and carry, and everybody had a conventional baler to make these. Then things got bigger, farms became more mechanized and small bales were seen as time-consuming and even a bit olde-worlde twee. Bales became super-sized, balers too, and the little balers were pretty much consigned to the scrap heap – or to hill farms. Our baler is ancient and is seriously temperamental. It doesn't like baling anything other than the driest of hay, clogging up the moment it encounters any damp hay. There's nothing digital or electronic involved; everything is about chains and cogs, with a mechanical knotter to tie the bales using old-fashioned sisal twine.

It is not a case of *if* something breaks, it's *when*. We have a ready supply of shear bolts which are designed to break when the baler clogs, to prevent the whole machine imploding. But real problems begin when something more complicated breaks. Spare parts can sometimes be found on Metal Mickey's scrap heap, and occasionally we've driven past a field and seen a baler like ours abandoned in a bed of nettles, and the farmer has agreed to us robbing it for bits. We have tried to order parts from an agricultural machinery dealer, but it's tricky.

'I need a bit for mi baler,' says Clive – and then there is a long-winded discussion about which bit he means, as we won't know what it's called. Whether our baler ever came with a manual I don't know, but even if it did it would have been written in hieroglyphics.

'And what make and model is it?' asks the young lad on the phone.

'A McCormick International B47.'

There's a sigh at the other end of the line.

'Could tek us a while to get that bit,' he says, and you might even catch a hint of a snigger in his voice.

Our hay-making machines are only used once a year for a few days, so it's hardly surprising that they are not exactly reliable. We now have a back-up round baler for when the weather does not allow us to get the hay dry or if the conventional baler gives up the ghost once and for all. Round silage bales are not ideal as they are not easily transported to our outlying heafs of sheep, but they are better than poor-quality damp hay. We learned about the perils of damp hay the hard way, watching our harvest go up in smoke after the stack spontaneously combusted.

No two years are the same. Sometimes we struggle all summer with changeable weather, and other years the barometer is set fair for weeks and we make hay under clear blue skies with no worry about rain.

The children love to be involved, and there's no shortage of jobs. Raven likes to do the lunch rounds with one of the horses, attaching panniers to the rear of the saddle and loading them with sandwiches and flasks of tea. Depending on which fields we have cut she may end up doing a large circuit, but her horse, Josie or Meg, won't mind because they can sample the goods along the way with a mouthful of hay here and there.

Reuben can usually be found mending something, or running back and forth to the tool shed for spanners and spare haybob tines. Miles, happiest when farming, rakes the hay. On blistering hot days you'll see him as a lone faraway dot, only distinguishable by his red bandana. Slowly he makes his way round the perimeter of the field, raking back the cut grass from the edges. This grass can't be reached by the haybob, and would be wasted if it were not for Miles and his rake. Others take their turn with the rake, but they don't have his diligence. Edith, Violet, Annas and Sidney play amongst the drying hay, burying each other or sometimes making 'pikes' and 'cocks', sweeping the hay into big or smaller piles with their hands and then redistributing it when the fancy takes them. The children love running through and rolling around in the puffed-up, aerated hay, throwing it in the air and then whooping with joy as it showers down on them and Pippen the terrier, who shadows them everywhere.

The downside is that they soon encounter the dreaded klegs,

which can bite before you've spotted them. I was bitten by one on the top lip only a few days after having Annas, when I was hanging the washing out on the line. My lip started to tingle and swell and within half an hour I had a trout pout like no other, and as my milk had just come in I had a pair of rock-hard gravity-defying boobs to complete the 'cosmetic surgery gone wrong' look. The only compensation is that the kleg becomes very sleepy while feeding on your blood, and with one slap you can squish the little blighter.

The children's next port of call after the hayfields is the beck. The bigger ones plunge into the pool below the packhorse bridge, while the little ones paddle at the water's edge or wade in until the water laps at the hems of their sundresses. During a dry spell when the river is shallow, the usually well-camouflaged tiny trout are clearly visible below the surface, facing upstream in the gentle current. They lie very still, only coming to life and rising when dusk falls and there's a swarm of flies to tempt them. The children try to catch one, but the fish are too quick and dart away into the shadows of the riverbank. The nearest they've ever come was one year when Miles found a dead one. He was mighty pleased with his find, and he soon had a crowd around him, all vying to hold it and caress its scaly body.

'Look at the pretty orange spots,' I said. 'What a beauty, such a shame he's dead.'

'Can we eat him?' asked Edith.

'He's tiny, yer wouldn't 'ave to be so 'ungry,' I said. 'An' yer don't knaw what ailed him.'

'We'll bury him, then,' Edith decided.

I watched as Miles led the barefooted funeral procession across the rocks to a sandy patch, then the digging began. Down on their hands and knees, they scooped out the sand to make a grave for the fish, which they named Nemo. In the meantime the corpse lay in state on a small flat stone, awaiting the committal. It was nearly time for the ceremony when the funeral was gatecrashed by Pippen, who had been watching from a distance, waiting to pounce when the children's backs were turned. She deftly picked up Nemo and set off at speed with the fish in her mouth. There was a huge commotion when the theft was discovered. Pippen realized that her fish supper was in jeopardy, so she swallowed it whole to shrieks of horror from the children.

They didn't have long to mourn. The familiar clanking sound of the baler meant that everybody, both big and small, had to head back to the hayfields, where the trailers needed loading in order to get the crop in. As the tractor is driven around the field at a snail's pace, we walk alongside picking up the bales and putting them up onto the large flat trailer. The smaller children hitch a ride in the pickup, into the back of which any misshapen, loose-stringed bales are thrown and taken to be rebaled.

Having spent the previous couple of weeks clipping the sheep, my hands are soft from the lanolin in the fleeces, and this means that the baler twine digs into my fingers when I lift the bales. The winding handle on the top of the baler can be slackened or tightened to determine how compact the bales are. There is a happy medium when it comes to the weight of a bale. Make them too light, and when you take them out to the sheep in the winter you'll find that there's not enough to feed them; make them too heavy and you'll struggle to even get them onto the trailer and into the barn.

We don't bale until the middle of the day, when the sun is at its hottest. If the heat is on, we'll keep going all through the afternoon and into the evening, but we always have to stop before it gets too late. For even in the best summer weather, a dew settles, the hay becomes heavy with water and then you have to wait until the next day for it to be dry again. 'Could do nowt till dinner time 'cos there was a gay bit of watter on,' is a refrain you hear quite a lot round here. We start leading the bales down in the evening, hopefully getting the fields clear by bedtime. If any are left out overnight they will weigh a ton by the next morning, having drawn dampness up from the ground.

Then it's a matter of mewing, or stacking, it. This is crucial: you need to get it right and pack the bales in as tight as possible. The more compact they are, the better they keep, though the bales that touch the beams or the stone walls will always spoil a bit, which can't be helped. A human chain manhandles the bales off the trailers and towards the elevator, which is a sort of conveyor belt used to take bales of hay from ground level up to the top of the mew. We have a love–hate relationship with elevators: they are only as good as when they work. They are needed for about one week every year, so you wouldn't think that was a big ask. We had an engine-driven one that was a pig to start, and had a chain that came off its track with annoying frequency. One hay time it jammed after a hay bale became wedged between the rungs at the top, and pressure started building up – not just in the elevator.

'Stop it, Mand!' Clive shouted from the top of the hay mew. 'Press the bloody stopper.'

'No, I'll not,' I shouted back angrily, cowering at the back of the hay trailer. 'That chain's gonna snap and I don't wanna be near it when it does.'

Smoke began to billow from the engine housing.

'Shit,' said Clive.

'Water – I'm gonna get water, it's gonna catch fire,' I shouted. One large bucket of water later, and that was the end of the elevator.

Next came an electric plug-in elevator: worked all right, not temperamental, but also not very powerful. We could only put one bale on at a time, or the chain would stop moving and the bales would slide back down. The main issue, though, was: how many outlying barns have a three-pin power socket?

After the very sad death in 2003 of our dear neighbour Clifford Harker, who lived and farmed at Pry House with his wife Jenny, there was a farm sale of all his equipment. Clifford was a stickler for keeping all his machinery in immaculate condition. When he finished using a particular machine for that year, he cleaned it, oiled it, and then winched it up to the top of his barn so that it was in perfect order next time he wanted it. We never borrowed Clifford's equipment, because it was always in its original state, mint showroom condition, unlike anything we have at Ravenseat. We knew that if we borrowed it we'd bust it, dent it or ding it. His haybob had every single one of its original tines intact.

After his death Jenny asked us to look after Clifford's stock trailer, which was quite new and perfect. They're a target for thieves, who simply reverse up, couple up and drive away, sometimes even loading them up with quad bikes, generators and anything else they find in the yard. Ravenseat is more off the beaten track, and Jenny felt it would be safer with us. She should have known better . . .

It was a standard-size double-decker sheep trailer, and we parked it out of harm's way at the top of our farmyard. The handbrakes on our trailers never work: they tend to seize up, and nobody gets round to repairing them. So it never crossed my mind to put the brake on Clifford's trailer. A few nights later we knew a storm was brewing, a heck of a wind was picking up, and we went out to batten down the hatches. The upside of a clashy night is that the draught makes the fire draw nicely, so Clive and I were all set, curtains shut, snuggled up on the sofa in front of a roaring hot fire. The television was out of action, as the wind was moving the satellite dish out of alignment. It was very late, we were sleepy and about to go to bed when there was an almighty crash from outside.

'What in God's name was that?' I said, pulling on my coat that was hanging from the beam above.

'I'll look after things 'ere,' said Clive, not moving.

I pulled on my wellies, switched the light in the yard on and opened the door. There were buckets scattered across the yard; a pair of stinking waterproof leggings that had been hanging on a gate to air were now lying in a crumpled, wet heap by the wall. Anything that hadn't been anchored down was strewn across the yard. This included one rather big thing that had not been anchored: Clifford's trailer.

As my eyes adjusted to the darkness away from the farmhouse I saw, at the far end of the yard, the rear end of the trailer. The wind had set it moving, the gradual slope of the yard had increased its speed, and it had come to a halt against the garth wall, the front of the trailer crunched and the drystone wall down.

We were mortified, but Jenny is so kind and gentle, she just said: 'Oh, Clive . . .'

When it came to the farm sale, everything was immaculate apart from this crunched trailer. 'Lot number 325,' the auctioneer shouted. 'New Ifor Williams trailer, cosmetic damage to the front, sold as seen. Grab yerselves a bargain.'

Clive was at the sale, and he went off with strict instructions: 'There's only ya' thing yer need to buy, an' that's Clifford's elevator.' It was ancient, but of course brilliantly maintained, and unlike ours, reliable. It really would be a godsend at hay time.

When Clive drove back into the yard, his mood was flat. Not unexpected: farm dispersal sales are never a happy occasion. There is a feeling you are prying and picking through someone else's possessions, accumulated over a lifetime, and it is the end of a chapter. It feels uncomfortable, and sad.

'They were sellin' all t'lal' stuff an' I went to get a cup of tea from t'burger van an' when I got back it'd bin sold,' he said. He hadn't expected it to go through the auction so early: most farm sales have hours of selling boxes of sundry bits and pieces before they get to the bigger items.

He has never been allowed to forget it. It's not just me who says it; everyone who is up here at hay time, struggling to load the bales at the top of the barn, repeats it, like a litany: 'Yer should 'ave got Clifford's elevator.'

There is a sense of jubilation, a feeling of immense satisfaction, when the last hay bale has been brought in from the field and mewed into the barn. Hay in the barn is like money in the bank: it is the insurance that we need for the dark winter months when the ground is thick with snow and there is not a bite of grass to eat, when life is hard for both man and beast.

We have a small celebration after hay time, everyone tucking into a good meal, usually outside on the picnic tables. A collection of people who have been involved in the process with bronzed arms, blistered hands and the children with freckled faces all join together to eat and drink while the smell of warm hay on the breeze reminds us of the efforts of the previous weeks.

A couple of years ago, I noticed during our little harvest festival that Sidney didn't seem to be his normal chirpy self. He's a stoical, determined little chap normally, but on this occasion seemed quiet. Maybe he was just tired, I thought, or perhaps the sun had got to him. Nothing that a good night's sleep wouldn't put right.

He spent the next few days pottering around the farm, playing with the other children, but I could tell that something wasn't quite right. I made up my mind that if he didn't rally over the weekend, I'd have him to the doctor's surgery on the Monday. Sunday was a wet day; the children were confined to barracks and feeling the cold after their days out in the sunshine. As the rain poured, they played quietly, drawing and reading. Sidney was still very withdrawn. I picked him up, cradling him in my arms.

'Oh, Sidney, what's up?' I said. 'What's bothering yer?'

He said nothing, but he is never particularly vocal. He put his head back into the crook of my arm and that's when I saw it: something bright orange right up his left nostril. I walked towards the window to get a better look.

'Ga' an' get yer father,' I said to Miles. Clive was studying a show catalogue in the kitchen.

'Tell 'im to bring 'is glasses . . . and a torch.'

By now the children had crowded round to have a look. The foreign body was neon orange, spherical, smaller than a marble and high enough to be invisible without tipping Sidney's head back. But, thankfully, it had not gone far enough to slip down his throat or windpipe. Clive had a look, glasses perched on the end of his nose.

'Do yer know owt about this?' he said, glancing at Reuben, who was shining the torch up Sidney's nose. Sidney stayed still, occasionally blinking.

'It . . . erm . . . looks like one o' them beads frae tha' necklace,' Reuben said. He was right: the girls had broken a necklace a few days earlier, and multicoloured beads had showered the living room floor, bouncing this way and that and rolling under the sofa. I'd reckoned I had hoovered them all up . . . but clearly not quite.

'We're gonna 'ave to tek 'im to t'hospital,' Clive said, taking off his glasses and rubbing his forehead.

'I wonder if we could suck it out with a syringe,' I said.

'We could give it a try,' said Clive.

I laid Sidney on the window seat and fetched a new syringe – we have a supply of them for injecting animals. Sidney stayed still while Clive shone the torch, and I tried to suck the bead out. It was a doomed mission, as I couldn't get any suction because there wasn't a flat surface on the bead.

'Do you know what?' said Clive. 'I reckon yer could work it down from t'outside o' 'is snout wi' yer finger.'

Sitting Sidney up, I started by nipping his nostril on the bridge of his nose. Sure enough, with a little manipulation, the bead was dislodged and dropped out onto Sidney's lap.

'There, Sidney, look at that,' I said, picking it up and rolling it around on the palm of my hand. ''Ow long 'as that been up there?'

'A gay while by t'look o' it,' muttered Clive, his eyes on the dirty, snot-encrusted bead.

Sidney looked at it, picked it up with his chubby little fingers, clambered down from the window-seat-cum-operating-table and threw it with some force onto the fire.

I felt really bad that I hadn't noticed it before. But he'd given no indication that there was anything up there, and I can only guess how it got there in the first place. Not long before this incident we'd had another foreign object wedged in a nostril, but on this occasion it was Violet who had a yow roll up her nose. It was a lot more obvious because it was slightly protruding. Yow rolls, made of a mixture of compressed cereals, like sugar beet, smell delicious, but smells can be deceiving and they taste awful. Violet had been sniffing the rolls, maybe a bit too enthusiastically, and had hoovered one up. She was busy extricating it herself when I saw her. It came out easily enough, but we had a conversation about animal food being for animals and not for consumption or inhalation by her.

We make all our own hay and silage, but buy in animal feed and straw. We use the straw to bed up the barns for the cattle, horses and any sheep that are inside in the winter. There are different types of straw: pea, wheat, rape and barley, which is the one we choose because it is more absorbent. We usually buy the straw from a hay and straw dealer called Edwin, who buys it off the field from the arable farmers who bale it and load it directly onto his lorry. Living remotely, it's often the haulage costs that are the biggest expense but it's better to buy it at harvest time and fill the barn, rather than wait, as prices can increase as the season goes on. When the weather is good and we are hay-timing, the arable farmers are also harvesting their crops, so we usually get a phone call at lunchtime from Edwin to say that there's a load of straw coming our way if we're interested.

Edwin is so busy at this time that he calls in other hauliers to help out with deliveries. One glorious summer day a few years ago, one of these hauliers was given the job of delivering a load of big square bales of straw to Ravenseat. We'd met the driver before, a chap called Reg. He was a big hulk of a man, tall and strong, with thick-rimmed glasses, green wellies, a collarless shirt and an old suit jacket, topped off with a very battered straw Panama hat. He wasn't in the first flush of youth and neither was his lorry, an ancient, faded maroon Foden with 'Reg, the king of hay and straw' emblazoned across the front in gilt lettering. We always saw it approaching – or rather, we saw the straw approaching, as the lorry was almost completely hidden under its colossal and precarious-looking load. Reg didn't do half measures: when he brought you a load of straw, it was always a load and a half.

We'd ordered straw Hesston bales – the big oblong bales about three feet deep, four feet wide and eight feet long, terribly heavy, each weighing about half a ton but relatively easy to stack tightly in the big barn. They are good to split, with layers of straw coming away in canches that can be carried or moved around to the stables in the wheelbarrow.

Reg slowly brought the fully laden lorry into the farmyard, and parked up. Climbing down from the driver's seat, he took off the straw hat, lobbed it into the cab and mopped his brow with a cotton hanky. 'Warm enough for yer?' he said.

'Grand,' said Clive, appearing from the barn with Robert, who was helping out that day.

'I'm gonna get t'kettle on,' I said, ushering the children into the kitchen. 'Looks good stuff, by the way.'

'Aye, it's as dry as snuff,' he said, starting to loosen the first of the ratchet straps that held the bales on.

Clive and Robert both stood at the side of the lorry releasing more of the ratchet straps, talking with Reg about the summer, the crop and the weather, and taking time to coil the straps up neatly so they fitted snugly into a box on the side of the lorry.

I went back into the kitchen and started to assemble a tea tray, pouring glasses of cordial for the children and rummaging through a few tins for cakes and biscuits. The kitchen door was open, the children were squabbling among themselves and as I stood chiding them with my back to the door, I heard a muffled thump. I felt the reverberation through the kitchen floor, but I wasn't alarmed: we usually push the bales off the lorry with the skid-steer loader, then stack them in the barn when the lorry has gone. Then I heard Clive shouting. Turning and going to the door, I saw the bales from the right-hand side of the lorry were now strewn across the yard, but there was no sign of the loader – and there was no sign of Reg. Clive was bent double and yelling at Robert, who was standing, quite dazed, amongst the scattered bales. It didn't take long to work out what had happened. I stood at the door open-mouthed while Clive and Robert scurried between the bales, shouting: 'Reg, Reg, where are yer?'

They were moving bales like they weighed nothing, shoving them over and pushing them aside, until eventually they heard the stifled groans of Reg from under a bale.

''E's under this 'un,' shouted Clive to Robert, as they frantically moved the bales that had pinned Reg to the ground. Finally the last bale was moved to uncover Reg, who was in a terrible state.

Clive hollered to me to ring for an ambulance and then bring some towels. Reg clearly had a head injury: blood was pouring down his face. As he'd been smashed to the ground his scalp had been peeled right back by the roughness of the concrete in the yard, and we could see his skull ('degloved' is the technical term I later learned). His face was awash with blood and the obvious thing to do was to push the skin back into place and hold a tea towel over it. This seemed to work, quelling the worst of the bleeding. I tried not to look at it. Reg was conscious, and anxious to be up on his feet. But any movement made him scream out in pain. It was clear his injuries included broken bones.

'Lie still, Reg,' Clive said. 'There's an ambulance on t'way.'

'Gimme mi ignition keys,' demanded Reg. 'I wanna go 'ome, I don't need no ambulance.'

He tried again to get up, but the searing pain forced him back down, and I think he finally accepted that he was going nowhere without the ambulance. All we could do was wait. We talked . . . a lot . . . about all sorts, anything to take his mind off the situation and hopefully keep him conscious. How long did your mother live, Reg? How's the crop this year? Have you any brothers? We covered him with a blanket, because we knew from experience that it takes a long time for an ambulance to reach Ravenseat.

'What 'appened, Reg?' asked Clive gently. 'It didn't look to be leaning.'

'Nivver known that to 'ave 'appened before,' said Reg. 'The whole side fell off.'

'One of 'em 'it me in mi back,' said Clive. 'Sent mi flyin'.'

One of the top bales had hit Clive hard enough to propel him onto the grassy bank at the side of the yard, but thankfully had not actually landed on him. Robert had been standing nearer the lorry cab and had been out of harm's way, although he was shocked. I kept going backwards and forwards between the house and Reg, having put the television on for the children to distract them from what was going on outside. Clive was on nursing duty and told me: 'He says he'd like a drink o' watter.'

'Noooo, he can't 'ave one,' I said. 'I think they'll be wanting to operate on him as soon as they can.'

I was so relieved when I heard the ambulance siren drawing near. It had been a very long forty minutes.

'Now then, what's been going on 'ere?' said the paramedic. Clive explained while Reg insisted that he just needed patching up and then he'd be on his way.

'Yer not going anywhere yet, Reg,' said the paramedic in a matter-of-fact fashion. 'Yer going to have to lay still whilst we slide this board under you, and only then are you going to go anywhere . . . and that's to the hospital.'

Manoeuvring him onto the board was clearly a painful procedure. Reg was swearing and ranting whilst Clive tried to keep him calm. Once he was loaded into the ambulance I went to say goodbye and ask him if I needed to ring anybody to let them know what had happened.

'Noooo, don't tell our Mavis,' he said. 'She'll do 'er nut.' He started getting tetchy again.

The paramedic turned to me, frowning: 'Just stay there an' talk to Reg for a minute whilst I take his wellies off,' he said.

I thought he was going to pull them off, but he reached for a pair of scissors.

'Whaat yer doin'?' squawked Reg. 'Don't cut mi wellies.'

Even I could see that the wellies were cheap ones that set you back a tenner at a garden centre, but Reg was adamant.

'Oi, I only bought 'em last week.'

Nobody was listening. The wellies were cut off. One was full of blood, and through his sock poked a glistening white ankle bone. That was as much as I could take, and I scarpered. As the rear door of the ambulance slammed I could still hear Reg's shouts and complaints.

'We'll look after thi lorry,' shouted Clive, as the ambulance pulled away down our road.

'Clive, I'm goin' inside,' I said. 'I's not squeamish or owt, but I really need yer to pick them teeth up an' wash t'blood away, I just can't look at it.'

The teeth, luckily, were just Reg's dentures. Clive and Robert moved the bales into the barn, and then Clive drove the lorry down to the car park at the other side of the bridge. The events of the afternoon had shaken us, and when everything was back in order we sat down to talk about what had happened, and how lucky we'd been that nobody had lost their life.

Reg was in hospital for a very long time: both his legs were broken, and he spent a long time in a wheelchair. Even in hospital he still worked as a haulier, delivering precarious loads of chocolates, newspapers and magazines from the hospital shop to the other patients on the ward.

It was winter when Reg finally returned to Ravenseat for his lorry, limping back into the farmyard after being dropped off by a friend. His old lorry was not keen on starting, having been idle for so many months. He sat in the cab, his straw Panama still propped up on the dashboard and his empty bait box on the seat, while Clive and Reuben rigged up our battery charger. Eventually they managed to get her started, belching out smoke as Reg revved her engine. He wound down his window to bid us farewell. That was the last time he visited Ravenseat. There was no way that we could ever risk an accident like that again. We decided that from then on we would have all our straw dropped off at Kirkby Stephen, and would bring it up to Ravenseat ourselves in small loads with the tractor and trailer. Occasionally we see Reg out and about taking hay and straw for delivery – same old lorry, and he still looks like he's carrying twice as much as anybody else.

Reuben was the right choice to help Clive jump-start the lorry, because he's a natural mechanic. He loves tinkering with the tractors, and if a wheel on the quad bike or trailer needs changing, we call for Reuben. He has an old tractor of his own, given to him by Colin and Anne, farmer friends of ours from Weardale. He keeps it inside a barn for most of the year, but at hay time it comes out, does a few laps of the field and then invariably breaks down.

The tractor, a 1960 David Brown, had belonged to Anne's Uncle John, who bought it new from the factory in Huddersfield. Whenever we went over to visit their farm Reuben would make a beeline for the tractor, which was falling into disrepair in an outbuilding.

'One day, mi' lad, that tractor'll be yours,' they said.

And they were true to their word. To Reuben's delight they rang up one day and announced that the tractor was his, but with the proviso that they needed to borrow it back in a year's time to transport one of their friends to her wedding at the local church. The betrothed couple are vintage tractor enthusiasts, so it seemed more fitting to arrive at the church on a trailer pulled by a David Brown than in a Rolls-Royce. By that time it obviously needed to look better than it currently did. So that was the deal: Reubs would get it tidied up for the wedding, and then it would be his.

Reuben was terribly excited, and couldn't wait to go and pick it up from Weardale and start work on it. After days and days of pestering, Clive and his friend Fencer set off one Saturday morning with a low loader trailer to bring it back to Ravenseat. Of course it wouldn't start; no one expected it to. The plan was to roll it down the hill to get the engine started so that it could be driven up the ramps and onto the trailer. Clive sat in the driver's seat with the clutch depressed, and all he needed to do was let the clutch out when it was going sufficiently fast, with the hope this would turn the engine over.

The tractor was shoved, and soon gained enough momentum for Clive to let the clutch out. But nothing happened. The tractor was bowling on, gathering speed. Only then did Clive notice that the lever connecting the clutch to the engine had seized, and needed a kick to make it engage. The tractor was now going at a rate of knots as Clive kicked the lever repeatedly, to no avail. He decided that it was time to abort the mission, but was unable to get his thick-soled thermal welly back across and onto the brake pedal. He said afterwards that he wasn't worried, but his face told a different story as one wheel of the speeding tractor went up onto the grass verge and the tractor began to bounce. The small crowd of onlookers lost sight of him as he rounded the corner at the bottom of the track, and Reuben said afterwards that he was waiting for the sound of crunching metal. It never came. Clive had the presence of mind to pull on the ratchet handbrake, and came to a very abrupt halt before he got to the main road.

After towing the tractor back to the top of the track they resorted to the traditional cure-all: a hammer. It worked – the hammer released the lever, and after a re-run the tractor started and was driven onto the trailer.

Reuben used some of his savings to buy new parts. We bought him an exhaust pipe for his birthday, some of the rustier bits were shot-blasted and the bonnet was resprayed red, with the wheel hubs in primrose yellow. He also made a little bit of money to reinvest in the tractor by swapping the spare double wheels that it came with for a pair of calves that he later sold. Finally, the tractor went back for the wedding. Reuben got an invite, but the downside for him was that he, too, had to undergo a transformation from an oily Fred Dibnah wannabe to a relatively clean wedding guest.

It wasn't long before Reuben gained another acquisition: a mini dumper truck. It had been languishing in a lean-to at our neighbour Totty's old house, up at Smithyholme, and for years Reuben had admired it from afar. After Totty died, the house and land was put up for sale, and Reuben was worried when he saw that the place was being cleared of all the rubbish and detritus accumulated by Totty and his mother. Some was being burnt, other stuff being put out for the scrap man.

'If yer want thi' dumper truck then thoo'd better do summat about it,' said Clive. 'Yer should put a note on it an' see what 'appens.'

'It's called being pro-active,' I said.

Reuben spent the afternoon writing his letter, even adding an

annotated diagram at the bottom showing his plans for the
dumper, how he was going to renovate it and what he was going
to use it for. He added our telephone number, slipped the letter
into a plastic sleeve and set off to Smithyholme, armed with a
roll of Sellotape.

We heard nothing for a while, and Reuben was getting dis-
heartened. Then we were called by Mr Smith, the executor of
the estate. The estate agent had been showing potential buyers
around, had seen the note and passed on the message to him.
Reuben talked with him at great length about the dumper,
which he had researched, explaining that it was a 1956 Rough-
rider. Mr Smith, either impressed or taken aback by Reuben's
enthusiasm, agreed that he could have it. All he had to do was
get it out of the lean-to.

And so began a labour of love. Almost every day Reuben
walked the mile or so to Smithyholme to tinker with the dumper,
trying to get it started. His luck was in when one day the estate
agent turned up for a viewing and found the dumper's starting
handle in the house. Having the starting handle made the whole
process of moving it much easier. By putting it in gear and
winding the handle it would move, fractionally. It was a very
long process. Every day he wound the handle and the dumper
slowly edged its way out of the lean-to, up the side of the house,
round a corner and then roughly about another thirty feet, until
it reached a place where we could load it onto a trailer and get
it home. Seeing the blisters on Reuben's hands from turning the
starting handle, we took pity on him, and on some days we all
went with him, taking it in turns to wind. It was a monumental
day when we finally got it back to Ravenseat. I thought that the
years of neglect meant that it would become another piece of
scrap adorning our farmyard – so I was confounded when,
within a couple of days, Reuben had it running. It comes in

handy, particularly in the winter, when I use it to move the muck out of the stables around to the midden.

Reuben has now progressed to a level where he actually fixes more things than he breaks. For many years we've had to tolerate him dismantling numerous broken items: pushbikes, lawn-mowers, washing machines, anything made of metal or with an engine. But recently he's had a few successes in the repairs department. Not long ago, one of our visitors to the shepherd's hut broke a wooden rocking chair. Admittedly, it was on its last legs and had been repaired a few times already.

'Chair in t'hut's knackered,' I said to Clive, as I came back into the kitchen to get the vacuum cleaner to finish cleaning the hut. 'I's thinkin' it needs slingin'.'

'Won't it glue?' said Clive.

'Nae, not this time,' I said. 'It's not safe to sit on. Where there's blame, there's a claim an' all that.'

We took it to the woodshed. 'I don't want any of mi walkers sittin' on it, neither,' I said as I writhed the armrests off, sweating and cursing, putting the bits on the woodpile. We snapped the spindles from the back, prised the curved rockers off the bottom and put a great deal of effort into breaking the chair into pieces that would fit on the fire.

When Reuben came home from school he disappeared off outside, as usual, only to emerge later looking particularly pleased with himself.

'Mam, I've got a surprise for yer,' he said, grinning from ear to ear. 'I've made yer a rockin' chair.'

In the summer the majority of the animals are out in the fields, which means that there are fewer daily bullocking-up routines. There may be a few baby calves in the buildings that need feeding, but on the whole the animals are outside, with us making daily visits to check on them.

There are plenty of other things to keep us occupied, as summer is when we have visitors to the farm: either overnight guests to the shepherd's hut, or daily walking visitors wanting cups of tea and home baking. The children are all good at looking after the visitors, even the little ones – but only once I'd trained them in customer relations.

At one time, when someone came to the door, they would open it a crack and mutter, 'Yeah, what d'yer want?' – now the door flies open and people are greeted with an effusive, 'Hello, how can I help you?'

Sometimes, though, they can be *too* good. One day it was pouring with rain, torrential, and Clive and I had been grumbling to each other all day about the weather. I was wet to my skin, as water had got inside my waterproofs. Clive had lost his hat, and he had a leaking welly.

'Ah can't wait t'dry mi arse in front of t'fire,' I said, cheering up at the sight of the smoke spiralling out of the chimney.

'Yeah, best moment of t'day,' he agreed.

When we finally got finished and back to the farmhouse, we found Reuben entertaining four wet walkers who were sitting on the fender in front of the fire, drinking mugs of tea.

'They're tired and wet,' he said, 'can't walk no further, they've phoned for a lift an' I've made 'em a cup o' tea . . .'

So while they hogged the fire, Clive and I hung around shivering and damp in the kitchen . . .

Throughout the year we buy in calves – Aberdeen Angus, British Blue and Friesians at about a week old, from a dairy farmer – and teach them how to lap milk from a bucket, rather than suck a bottle. Twice a day we mix powdered milk with warm water,

LEFT Clemmy, just hours old. RIGHT Clive, with Clemmy, out on the farm.

BELOW Clive and I still clip the old-fashioned way, starting from the back of the head.

Looking after the drystone walls around the farm is an endless project. It's all hands on deck when it comes to their maintenance.

There always has to be one!

River walks at sundown.

Our hay-making methods are very traditional. Nothing digital or electronic here!

It's a hay day, perfect for play.

Princess, Della, Josie and Little Joe on the moor.

ABOVE LEFT Violet and Fan watch over Queenie (with an impressive moustache).
ABOVE RIGHT Raven and Violet enjoy a quiet moment with veteran Meg.

Muker Silver Band playing hymns outside The Farmers Arms.

Driving the Fordson Major tractor back home from Muker Show.

We've used peat to colour this yow's fleece to enhance the whiteness on its legs and face.

ABOVE Raven in the pens at Muker Show. Sidney takes it easy on the bench.

RIGHT Clive showing one of his yows.

A line-up of Swaledale tups being judged at Tan Hill Show.

Mr Peacock helping himself to the remains of a cream tea.

Feeding yows on a winter's evening.

and after a few days of us putting our hands in the milk and letting the calves suck our fingers, they master the art of slurping it down.

One afternoon we had an American woman come by for a cream tea, and she talked to me about the animals. She had a languid American drawl which must have lulled me into stupefaction, because I heard myself saying, 'Do yer want mi to show yer around?'

I soon realized my mistake as we walked towards the barn and she told me that she was a lecturer in animal rights and welfare at a university in the States. But I knew we had nothing to hide, and that she'd soon be able to see the animals were all healthy and happy. I decided to start by showing her the calves that I was hand-rearing.

'But where are their mommas?' she said.

I explained that their mothers were all alive and well, but they were milk cows, and that a cow could either feed humans or a calf. In order to provide milk for human consumption the calf would remain with the mother for a few days feeding on colostrum and then be reared on a calf milk replacement powder, in the same way that a baby could be reared on formula milk.

'I'm their momma,' I said. She gave me a disapproving look.

'They tek no harm,' I said. 'In fact, they mek lovely cows. They're gentle wi' folks, as they've been 'andled.'

Perfectly on cue, the eight calves noticed that their human 'momma' had shown up and they came galloping towards me, anticipating a feed.

'Awwwww, can I touch one?' she said. I could see she was warming.

She scratched the top of the calves' heads as they jostled for position, all trying to suck her fingers.

'I feed 'em with a bottle for a start,' I said. 'An' then I learn 'em to lap out of a bucket: I put mi 'and in t'milk bucket, then they suck mi fingers. Look.'

I held out my hand to the calves, and sure enough one of them nuzzled me, and then started sucking my fingers.

I felt I'd won her over and while I was on a roll I kept talking, my back leaning against the barrier gate. I told her all about Ravenseat and our cows, pigs, horses and sheep. I wasn't taking much notice of the calf sucking my hand, only turning to look at it when I felt a tug on my finger. The little Aberdeen Angus calf stopped sucking, gulped, gave a wide yawn, licked his lips and then sauntered off to the other side of the pen.

The American lady was chattering away, but I wasn't listening. With my right hand I was feeling the wet, wrinkled ring finger of my left hand. Something was different about it . . . it was a naked finger. My wedding ring had gone.

I brought the conversation to an abrupt close. I needed my newfound American friend gone if I was going to wrestle the calf to the ground and try to get it to regurgitate my ring – I certainly wasn't going to do it in front of her. The best I could do, in the meantime, was make a mental note of which calf was the culprit.

I told Clive.

'Are yer sure?' he said. 'Was yer ring loose?'

I explained that it was looser than normal, due to clipping time and my skin being softer and smoother than usual.

'Which calf were it?' he asked.

I pointed to him, sitting in the straw, oblivious to the fact that he had swallowed an eighteen-carat symbol of Clive's undying love for me.

'Reeuuuuuuuuuben . . . ga an' get the metal detector,' he yelled across the yard.

The calf stayed still while Clive hovered the metal detector over it. It clanged. Reuben got excited; Miles, who was also now in the pen, was a little more dismissive. 'That's its bloody ear tag,' he said. 'They'll all clang if yer wave it over their head.'

'Yer reet, Miley,' I said. 'But dun't sweeear.'

It was a pointless exercise. Even if we located the ring, it wasn't going to help get it back. We decided to put the calf in solitary confinement until nature took its course. But the calf had other ideas. Alone in a stable, he bawled his head off and strained against the wooden hurdle, desperate to be reunited with his friends. We soon put him back.

Bribery was my next strategy.

'Twenty quid for whoever finds mi ring,' I said.

Every day Reuben methodically did a sweep of the calf pen, finding all manner of things, but no ring. After about a week I was giving up hope, deciding that the ring would stay in the bottom of one of the calf's stomachs forever. Then, late one afternoon, Reuben approached me with outstretched hands.

'Look 'ere, Mam,' he said, unable to conceal his glee.

There, in his grubby hands, was a dollop of muck. I raised my eyebrows and closely studied it, trying to look as interested as it's possible to be in a calf turd.

'Yer ring,' he said excitedly. 'It's in 'ere.'

He went over to the water tap and started to rinse his hands, keeping them tightly cupped together.

He was right: within a few minutes I was slipping my wedding ring back into position. Reuben's hands remained outstretched, only this time I crossed his palm with silver . . . or rather, a crisp twenty-pound note. Proving that, as the old saying goes: where there's muck, there's money.

Our cream tea visitors are sometimes a wonderful diversion from the routines of the farm. It was a glorious summer's day when I took a phone call from Ian, the verger at the church in Kirkby Stephen.

'Amanda,' he said. 'Can you do me a favour?'

'Go on,' I said.

He explained that the evening before, he had been in the church quietly preparing for vespers when a small group of people had wandered in. This was not unusual, as visitors often came in for quiet contemplation or to admire the medieval architecture and ancient tombs. Ian was sorting the hymn books when he heard the piano in the chancel being played. Brilliantly.

'Such talent,' he said to himself, stopping his work to listen.

The other three visitors were sitting in the pews while the fourth played. Ian was reluctant to disturb them, but eventually his curiosity got the better of him.

'Hello,' he said. 'Your friend plays beautifully.'

'G'day, mate,' said one of the three. 'We're from Sydney in Australia and mi mate here Duncan is a concert pianist. We're doing the coast-to-coast, and we've been walking for a week now. Duncan here was getting piano withdrawal symptoms, so we decided to check in to our B & B, drop off our rucksacks and then see if we couldn't find him a fix.'

'I tell yer what,' said Duncan, turning round from the piano. 'I wouldn't mind a go on the organ over there.' He nodded towards the church's substantial pipe organ.

According to Ian, what followed was a beautifully moving, impromptu organ recital.

After they left, the only way Ian could think to thank them was by asking me to provide them with cream teas and a go on our wonky, out-of-tune piano for their next stretch of the coast-to-coast journey.

So that is how I came to have a party of four Australian

walkers sitting in the living room, one of them perched at the piano playing a slightly off-key rendition of 'Moonlight Sonata'.

It is fair to say that when you visit Ravenseat, you must expect the unexpected. You never know who or what you'll find. Usually you'll be pleasantly surprised, but this was not the case for a couple of guests to the shepherd's hut.

One of our crusties, a really, really old yow, was in the garth next to the shepherd's hut when she reached the end of her natural life, at the grand old age of nine. She was certainly a creaking gate and had been looking very frail for a long while. I was reluctant to move her from where she felt at home, so I'd warn our visitors: 'Tek no notice of 'er. She's a real veteran, an' 'appy enough, she's going to die a nice, peaceful, natural death any day now. She's 'ad a good life.'

Next morning, when Clive and I were taking the guests' breakfast, we found her: she'd passed away in the night. Her bloated body was splayed right across the path next to the hut and she didn't look a bit peaceful: her face was contorted like something out of a horror movie.

'Ach, yer owd bugger,' I muttered to her. 'You've only done this to me because you know I've got people in the hut. After all we've done for you . . .'

9

September

September means the return to school, with the usual panic as sandwich boxes are fished out of schoolbags that were abandoned several weeks ago. Invariably, one of them will have a mouldy yoghurt pot lurking inside. School uniforms are dug out of cupboards, and with luck everybody will just move up a size, so that the jumper that fitted Edith will now fit Violet and in turn the jumper that fitted Miles will now fit Edith. Screwed-up sheets of homework are found, and someone will hurriedly write their 'what I did in the summer holidays' essay. I have sleepless nights worrying about this regular project and all the things that we didn't do. Didn't go on holiday, didn't go to the seaside, didn't go to the shops. Last year I only realized at the end of the holiday that out of eight children, five of them hadn't left the farm at all for the entire seven weeks. Helping Edith with her essay, I focused on the things they *had* done: digging in the sand on our 'beach' at the side of Whitsundale Beck, snorkelling in the dub below the Jenny Whalley waterfall, riding Little Joe and Josie bareback around the fields, hunting down our piggies when they went on the run after escaping from the graveyard. They slept in tents, cooked their own tea over an open fire and partied in a thunderstorm with a group of motorbike enthusiasts from Hull. I decided that they'd probably had a pretty good time.

I soon rectified the 'not leaving the farm' issue by taking them to Muker Show – our local show, held on the first Wednesday of September. Sometimes term time will have officially begun, but we still all go. There's a programme of events: a sheep show, sheepdog trials, tents full of produce and horticulture, a dry-stone walling demonstration, a vintage tractor and implements competitions, and quoits. There are lots of children's competi-tions too, with handwriting, drawing and photography classes. Our children will bake cakes, do a few drawings and enter the young handler and stock-judging classes.

Things are always frantic on the morning of the show. When the show schedules are sent out in early August it all sounds like it's going to be such jolly good fun, baking loaves of bread, making gingerbread, shortbread, biscuits, curd tarts, cakes, the dreaded scones and a Battenberg . . . It's only when we come to start the epic bake that I think properly about it all. A *Batten-berg*? What made me tick that box? In my world, only Mr Kipling makes them, or anyone with a lot more time on their hands than me. Trying to load all these goodies into the Land Rover with eight children, plus their artwork and everything a family needs for a day out, is a nightmare. Butterfly buns get sat on, doilies are forgotten – and then there's the sheep and tractors that also need to be transported.

Miles and Reuben like to run in the junior fell race, training all summer without even knowing it by coming to the moor to gather the sheep, running through the heather, up and down the hills. To take part they need trainers, shorts and a T-shirt. I had enough to do: *it's time to delegate*, I thought.

'Reet, you two, I want yer to pack yer runnin' gear, everything yer need for t'race,' I said, leaving them to it. Later, when we were at the show, the announcement came out over the tannoy:

'Entries now open for the junior fell race. All runners to assemble at the bottom tent.'

Quick as a flash, Miles and Reuben rushed off to the Land Rover to get changed and then went to put their names down on the list. They returned minutes later.

'We cannae run,' said Reuben, his eyes filling.

'We're not allowed,' added Miles, wiping tears from his cheeks.

'Stop thi' sobbin' an' tell mi why,' I said, although it wasn't hard to guess. Each boy had two pairs of trainers: a spanking new pair, bought specially for the start of the new school term, and a knackered pair that had been worn out over the previous school year. These were on the tight side, had seen better days and were due to go on the fire anytime soon. Typically, these were the trainers they'd brought with them. The old trainers were deemed unsafe for running the fells in. So much for delegating some responsibility to them: I now wished I'd overseen the packing of the running kit.

All was not lost: there was still a chance. The seniors did not run until later in the day, and as the boys have big feet, we wondered if we could borrow a couple of pairs of trainers.

'We'll put a request out over the tannoy,' said the show secretary.

Not only was I going to have the ritual humiliation of losing in all the baking classes, it was now going to be broadcast over a loudspeaker that my children didn't have a decent pair of shoes. We were lucky: a couple of lads from the team of local harriers were happy to lend their shoes to the boys.

Borrowed shoes were no hindrance to Reuben, who won the race. He was ecstatic, waving his trophy aloft in front of the cheering crowd with a delighted Miles, who also finished well, at his side. The lad who lent Reuben the shoes was also delighted: 'I've nivver won nowt, but at least mi shoes 'ave,' he said.

I love the show, with its unwavering traditions. The people may be different, the sheep may change, but year on year the show remains the same: the drystone walling demonstration, where the same gap in the wall is rebuilt every year, the chainsaw sculptor cutting out squirrels and mushrooms from logs, with Reuben pestering him until he hands over his offcuts for us to take home for our fire. Muker Silver Band plays the same tunes, while folk sit on the embankment eating ice creams and watching the dog trials taking place at the bottom of the field. One year I inadvertently sabotaged the dog running event with a half-eaten beefburger that one of the children couldn't finish. Getting out of the Land Rover, I casually lobbed it over the wall. It went further than I expected, landing by the dog trialling gates.

That'll test 'em, I thought.

The quoits competition takes place in the afternoon, when the sound of the band music is punctuated by the chink of the iron rings hitting one another. I enjoy teasing the players:

'Playing hoopla again?' I say.

'We're athletes, I'll tell ya,' one of them replies, indignantly, pint in one hand and quoit in the other.

'Where's Dave at anyway, is he not playing today?' I ask.

'Nah, off with a sporting injury, he's 'avin physio. It's a shame really 'cos we've qualified for the internationals . . .'

I'm impressed, never having realized that quoits are played on a world stage. 'Where do yer 'ave to go for that, then?' I say. 'A lang way?'

'Aye, we've gotta bus gaan, we're playin' at Whitby.'

'Crikey, that far,' I say, trying not to laugh. 'Who are yer up against anyway?'

'North York Moors league.'

There really is a truly international feel to it all.

The sheep show is solely for the Swaledale sheep purist, there are no other breed classes. To be the champion at Muker is a tremendous accolade, as this is the heart of Swaledale country. Competition is fierce, with a lot of effort going into preparing the sheep. Days are spent tonsing, colouring and trimming. There are many classes, and filling in the entry form is a challenge in itself. For example, there are large breeder and small breeder classes, district and open classes, gimmer lamb, gimmer shearling, gimmer shearling *and* gimmer lamb, best gimmer in the district, gimmer that's been shown but hasn't won a prize . . . and that's just some of the classes for females. It's in the tup classes that things get *really* complicated. We sometimes get our show entries wrong, embarrassing ourselves on the show field by taking our sheep out into the ring only to find that we haven't entered that class.

The whole sheep show takes all day to judge. Serious deliberations go on for hours, both judges and bystanders debating each sheep's qualities amongst themselves. The judges eventually whittle down the sheep eligible to compete for the overall championship. The Swaledale enthusiasts spend the whole day in the sheep pens, only moving to perhaps go and buy a bacon sandwich, and then it's straight back to where the action is. They discuss whose sheep are on form, who is looking well set to win and who is not showing that year, and why. Even after it's all over, with the champion sheep declared, the arguments continue well into the night in the Farmer's Arms pub.

Classes in the produce tent are also hotly contested. I'm already psychologically on the back foot, having seen the other entries. I'm unnerved by the neatly regimented scones, my spirits sinking like the middle of my gingerbread. I'm a stickler for the rules, eyeing up a rival entry in the 'two items from a Dales kitchen presented on a wooden board' class, noting that she's put a jar of home-made piccalilli alongside a fan of home-made digestive biscuits.

'Whhhhhat!!!! She's gonna get disqualified,' I say to Raven. My Yorkshire curd tart and oatcake would have looked better if I, too, had arranged my triangular oatcakes into a pattern, but I thought that I could only put two things on the board.

Returning after the judging, I see a little note on the other entry saying in big letters: 'Disqualified . . . TWO items please.'

Well, at least I can count – but I still don't win. The judges' decision is final, even if they don't know a good sheep/cake/vintage tractor when they see one.

My friend was sure she was going to be awarded first prize in her flower-arranging class, as nobody else had entered. Later on, she told me she'd been given a second, as her arrangement was not considered good enough to warrant a first. Firsts are only awarded at the judge's discretion.

The vintage tractor classes give us a logistical problem with getting them there. Reuben would dearly love to be at the wheel of his David Brown tractor but he's too young to drive on the roads. The night before the show Clive and I take it and one of our other classic tractors, complete with muck spreader on the back. It's about a five-mile journey from Ravenseat and at vintage tractor speed it's very pleasant on a lovely late summer evening, with time to look over the walls at the views that remain much unchanged since the times when life moved at a slower pace. We chug our way down the dale in the open-topped tractors, with the setting sun casting a glorious warm light over Kisdon. After parking on the show field, we wait for my friend Rachel to pick us up and take us back home.

One year our journey was rudely interrupted when an articulated lorry became wedged on the Usha Gap bridge. The supersized vehicle had tried to negotiate the small narrow bridge and was stuck, unable to move backwards or forwards. There was a tailback of other would-be exhibitors trying to get to the show field and there was no other route, not without doing at least a thirty-mile detour: not a practical solution for our elderly tractors. The driver couldn't have timed it any worse. He sat on the parapet of the bridge, his head in his hands while a procession of people who had now abandoned their vehicles filed past his stationary lorry.

'What was 'e thinkin' of?' they'd mutter, and then take pictures of the stricken lorry on their phones. We decided to leave our tractors in a nearby field overnight, and thankfully by the next morning the lorry had gone.

It is a surprisingly common sight, outsized trucks and tankers trying to weave their way down the narrow Dales roads. We blame the growing reliance on satnavs. The bends are so tight that something has to give, and usually that something is a drystone wall or two.

The second half of September usually brings a slowdown in our visitors. The number of coast-to-coast walkers dwindles and I revert to doing what I came to Ravenseat to do: shepherding the sheep. I live a life of two halves, a double life. During the summer I spend the mornings feeding the pigs, calves and the few other animals around the yard, and in the afternoons I provide refreshments for the visitors, fitting in whatever jobs need doing around the farm that day. If I'm working in the sheep pens or in the barn, I'll nip back and forth between the picnic benches, kitchen and sheep pens. During the summer months the world comes to me, and then through the winter I get the quietness and seclusion of a remote hill farm: the best of both worlds, I think.

'Have you ever walked the coast-to-coast?' visitors ask.

'Nah, but I reckon I walk the equivalent o' t'192 miles of it just going backads and forrads carrying cups o' tea.' Plus I feel that I know the route very well, having had the same conversations for a good many years now.

'Where are you going tomorrow?' I ask, making polite conversation.

'Reeth, I think,' they say, flicking through the pages of the Wainwright's Coast to Coast guide.

'Really? And then to Richmond,' I say.

At the end of the summer I try to add up how many walkers I've taken pity on and shipped to Keld, those who called it a day when they got to Ravenseat, asking us to phone for a taxi for them, or wanting directions to the nearest bus stop. We give them a lift to Keld, but they may have to wait until it's convenient. Occasionally, when they look as though they will expire from exhaustion, I drop everything and take them immediately. One rainy afternoon Raven took a tray of hot chocolates down to a small party of walkers who were sitting in the woodshed, and reported back that one of them was covered in blood.

'What's 'appened?' I asked her.

'Fell over summat,' she said in an unworried way. But I *was* worried: 'where there's blame there's a claim' sprang to my mind again, as I thought of Reuben's inventions littering the woodshed. I went to investigate.

Sure enough, a young woman in a bloodstained cagoule was sitting dejectedly on the bench, her friends bombarding her with questions.

'How do yer feel?' they said. 'Do you feel dizzy? Sick?'

She shook her head. 'I'll be OK,' she said.

Yer snout's mebbe busted, I thought, but didn't say. Her nose looked lopsided.

'What did yer fall ower?' I asked, looking nervously around the woodshed for likely obstacles.

'She fell coming down the road,' said one of the group. The young woman nodded.

'Yer stick was tucked behind yer arms,' said another. She nodded again.

This meant she couldn't stretch her arms out to save herself.

'An' yer bootlaces was undone,' volunteered another. She nodded yet again, miserably.

'Was there a banana skin, too?' I said, then remembered that it wasn't funny.

I took the whole group to Keld, to the B & B where they were staying.

I was reminded of Jimmy, our old friend and neighbouring farmer. Jimmy was driving along the road in his large blue tractor, towing a stock trailer on one wet and blustery afternoon. He was flagged down by a group of bedraggled, weatherbeaten hikers who begged a lift. Jimmy, being the kind and generous person that he was, agreed that they could get into the trailer and he'd take them the three miles or so to Keld, which is where he was heading. After shutting the jockey door on the trailer he climbed back into the tractor driving seat and set off again, homeward bound. Like most farmers, he was planning in his head the next job that needed doing, which distracted him, and he completely forgot that he had passengers. He took a detour, off the road, down a track and into the Greendale, a steepish field in which his draught yows were grazing. He saw one was hotchen, quite lame. Having no dog with him, he put the tractor into a higher gear, put his welly on the accelerator and gave it some gas across the field, the trailer rocking from side to side behind. The sheep flew to the bottom of the field, with the tractor tearing after them. Only when he had them corralled in the makeshift hurdle pen did he switch the tractor engine off.

Loud groans and gasps were coming from the trailer. He thought for a moment, then it all came flooding back. He'd picked up a group of walkers, hadn't he? He remembered offering them a lift to Keld in the trailer.

Apologetically he opened the trailer door, and the ramblers tumbled out.

'I's proper sorry aboot that,' he said in the broadest of Yorkshire accents. 'I'd forgetten that thoo was in t'back.'

Dusting himself down, the leader of the walk muttered something, gave an almost imperceptible nod, then said, 'Yer know what, I think we'll walk from 'ere.'

As anyone who has ever travelled in a trailer knows, it's unpleasant at the best of times – not to mention illegal. There have been times when I've been crouched in a trailer, holding a tup's horn with the explicit instruction from Clive, 'Don't let the beggar rub on owt.' The aluminium insides of a trailer can sometimes stain and spoil the hair on a tup's face, giving it a greenish-grey tinge. This is unforgivable at the tup sales. When we get there, Clive's tup exits the trailer with his hair intact, but Clive's wife looks a little green about the gills after a rough ride across the Buttertubs.

Back in the days when I was a contract shepherdess, I would often drive around in vehicles that weren't mine. On one occasion I was asked to pick up some fence posts and wire from the agricultural suppliers. I borrowed the farmer's beat-up Astra van, taking it home with me, having arranged to return with the stuff the next morning. I was secretly pleased about this arrangement, as it had a full tank of diesel (probably red, but we won't dwell on that).

My ancient pickup truck was a thirsty old thing and I was always counting coppers to raise the money for the minimum delivery of diesel at the petrol pumps, so the Astra van was a real treat. For one night only, I could go places!

Actually, I didn't want to go anywhere much – but there was one job that I needed to do. One of my other employers, Pat Bentley, an alpaca farmer, had kindly offered me the chance to graze one of her fields with my mini flock of sheep – the pet lambs I'd been given and had hand-reared on the bottle. There were Connie and George, the Swaledales; Peter and Paul, the Suffolk crosses; and the mules, who were affectionately known as the princesses. Flymo the goat had not been included in the invitation, as her reputation as a ratcher and wanderer went before her. She was staying at home, which was just as well, as there was little room once the sheep had been tempted into the back of the Astra. It was standing room only in there, so I hastily shut the double doors before jumping into the driver's seat and setting off. I drove slowly, suspecting that I was seriously overloaded as the brakes didn't seem to be working so well. The noise was deafening, the bleats reverberating around the inside of the van, sounding as if they were in a tin can. There was no mesh grille between the cargo area of the van and the cab, so I wasn't surprised when one of the princesses appeared and rested her head on my left shoulder.

'Yer a softy,' I said, glancing at her. Her dark eyes stared fixedly ahead, so bright and clear that I could see a reflection of the world passing by the window. I drove along back country lanes and then a short distance along a main road before a right-hand turn off to Newby. I soon backed up to the field and let them go. They hardly moved away from the gate, putting their heads down to graze straight away, pleased to be out of the cramped van and delighted with the lush grass. That was it: job done. I tied the gate and set off back home, already thinking about the valet required in the van back. A bucket and brush would clean it up sufficiently, as it had not exactly been spotless before the sheep-shifting job. The following day I turned up at work in the van with the items needed for the fencing job.

'All reet?' said Mike, the farmer I was working for.

'Yep,' I replied.

'Do yer knaw owd Willy?' he said. 'I seed 'im last neet, an' I sweear 'e's getting a bit too much pop these days.'

He told me that Willy, the odd job man who knocked about the village and was often found propping up the bar in the Sun Inn, had been telling a tale that he'd been walling a gap along the main road and had seen a van being driven by a sheep.

'Honestly, I'm tellin' thi,' owd Willy had been saying to anyone who would listen. 'As plain as day I see'd a van, 'twas stopped, turnin' reet an' there was a mule in t'drivin' seat.'

'Oh, aye,' said the doubters.

'An' I'll tell thi summat else – it looked like thy van,' he said, wagging his finger at Mike.

September is the beginning of the 'harvest of the hills', when sheep off the moors and fells of northern England are gathered in and prepared for sale at the auction marts. The crop of lambs that are not required as replacements for the flock and the older yows (known as draughts) are sold in the coming weeks. It is now that the lambs are taken off their mothers (speaned) and the yows sent back to the moor to get into peak condition for the next breeding season.

The land that was haytimed earlier in the summer will now be flush with a regrowth of grass, and the lambs go into these fields. The change from grazing on sparse moorland to the lush green meadows gives them a lift, and makes them bloom ready for sale. It's important to get the lambs gathered in from the moor and weaned, because a male lamb left wandering among the yows can be a problem if he starts doing what comes naturally. All thoughts of carefully selecting which yows are going to which tups go out of the window if a nameless tup lamb gets there first.

The first sales are those of the mule gimmers. A mule is the result of crossing a Swaledale yow and a Blue Faced Leicester tup. They are a very popular breed, the females primarily used for the breeding of fat lambs. We dip them with golden bloom, which pearls their coats and turns it a pale biscuit colour, and we shave their necks and bellies to make them appear taller. Then their brown and white mottled faces and legs are washed and they are sorted into smaller groups: a pen of light ones, a pen of dark ones, a pen of strong ones and a pen of lesser ones. Once upon a time we tied strands of different-coloured wool into the fleeces to identify the different groups, but this has been out-lawed, and we now use small pots of coloured paint instead. Apparently the strands of wool stayed in the sheep's fleeces until clipping time, and when the freshly shorn wool was being pro-cessed, the 'foreign' wool strands contaminated the batch.

All sheep must now be electronically tagged with a yellow tag containing a microchip, so that they are traceable back to the holding of birth. When electronic tagging was introduced farmers were naturally suspicious about how it was going to be implemented and, more importantly, how it would work in the field. For us, breeding pedigree, it is a great asset. We have a hand-held electronic tag reader that tells us a sheep's parentage – and in some cases nowadays, the grand-parentage.

Before electronic tagging, we all assembled in the pens before tupping time, looking up each individual yow's pedigree and consulting a dog-eared flock book. It usually went like this:

'Alec, read this tag, would ya? I cannae see it.' Clive would have a yow caught in the corner of the pen and be rubbing its grubby ear tag with his finger. Alec would stride over and peer into the yow's lug too, then start fishing around in his pocket.

'Where's mi glasses?' he'd say, still rummaging. 'I's thinkin' it's 866 – nope, it's mebbe 998. Or is that 8 a 3?'

'Christ, I'll just read it myself,' I'd say. Then I wouldn't be able to find the number in the book, and when I did, I'd be faced with an illegible scrawl where the sire had been hurriedly written down. Then it would start to rain . . .

Now we just point the scanner at the sheep's yellow tag and we find everything we need to know: date of birth, parentage and progeny. The only way the scanner could be improved would be to make it so sensitive it could read the tag of a yow moving at high speed up a field.

Clive has flock books going back many years, and I have been lucky enough to see some of the sheep records that belonged to my neighbour Rachel's forebears, which go back even further, with records of every sheep by name. The names were descriptive: *coppy legs* (a 'coppy' is a milking stool), *yan 'orn, lang tail*, for instance. The records included which heaf each sheep came off, and how it bred. Traceability may be a buzzword nowadays, but these ageing records show that it isn't a new thing.

I sometimes feel a touch sad that I personally do not have the same link with Swaledale that others do, to be able to say that I tread the same paths as my ancestors. But perhaps my longing to fit in to this landscape is what has spurred me on to write, record and absorb everything relating to Ravenseat that I possibly can. Traces of the past come to light all the time, small things of no real significance, but tangible evidence that this place has been a home to people for many centuries.

We were lucky enough to be given a copy of a map of Ravenseat from 1771 with the springs, wells and stiles all carefully marked, so the children and I went to one of the spots marked and on the hillside we uncovered a hollowed-out stone that once caught spring water as it bubbled out from the ground. We followed a path shown on the map from the farmhouse to the West End field, the route taking us to a drystone wall. Sure enough, now that our eyes were opened we found the stone stoops and the iron hangings from which a small gate once hung.

A fascinating and obscurely named book, *A Bonny Hubbleshoo* (dialect for 'a complete jumble') gives a unique insight into what life in this dale was like in the late nineteenth century. One of the passages, 'Rambles in Swaledale', was scathing about the area, and it's not surprising the author chose to remain anonymous.

Although in the matter of cleanliness the housewives of the working classes in Swaledale compare favourably so far as nattiness and general freedom from filth is concerned with the matrons of any other part of England, yet looking at the dwellings of the humbler classes suggested to us the thought that their mode of construction belonged to the dark ages . . . Their great want is ventilation. The rooms are too low and the windows too small . . . The window is seldom opened and rank odours are allowed to accumulate . . . So far as the outside is concerned, we noticed an attempt at external decoration which was not very artistic, giving them in a number of instances a grotesque appearance, suggesting a strangely heterogeneous combination of mingled barbarism and civilization . . . Probably one of the most unsightly and unhealthy appendages of the humble domicile of the Swaledale miners are the middens, or mounds of ashes and dung which are close to the door. Their peaked summits, constantly emitting a smoky vapour and a foul stench, poison the blood and render the neighbourhood very unhealthful.

I read it and looked out of my little kitchen window across to the muck midden in the yard, and smiled to myself.

Tup crowning day falls roughly in the middle of September, about the time of my birthday. Tup crowning definitely ranks as the more important of these two events, and as time passes I'm inclined to agree with more emphasis being placed on it. The secretary and two elected members of the Swaledale Sheep Breeders Association visit farms in the district that hope to sell pedigree Swaledale tups, to check that they meet the breed standards.

The Association was formed in 1920 and one of the founder members, Raper Whitehead, was farming at Ravenseat at that time. There have always been sheep roaming the moors but the association was set up to ensure the purity of the breed, and to set a standard for all breeders. The predecessor to the modern Swaledale sheep was described in 1794 by a writer called William Marshall:

The moreland breed of sheep has always been very different from that of the vale and has not varied, perhaps during a succession of centuries. It is peculiarly adapted to the extreme bleakness of the climature, and the extreme coarseness of the herbage. They live upon the open heaths the year round. Their food, heath, rushes and a few of the coarsest grasses, a pasture on which, perhaps, every other breed of this kingdom would starve.

Their horns are wide, the face black or mottled . . . their wool longer and much coarser than that of Norfolk sheep.

You can see in his description that these were the forebears of our Swaledales: we may have refined their looks a bit, but that essential hardiness is what they are about and what we, and the buyers at the auctions, want from them.

The procession of spectators following the crowners increases as they make their way up Swaledale, and as we are one of the last in the line, a cavalcade of trucks and Land Rovers follows them into our yard. Farmers want to see what other farmers' tup shearlings are like, because they are looking for new bloodlines to introduce to their flock. Each tup is presented to the inspectors, who are looking for obvious defects and checking that it's a strong, healthy animal. When they are satisfied that the tup fits the breed criteria the district secretary puts a tag in his ear with the association crown on one side and his official registration number on the other. Once upon a time, the crown was burned into the horn with a hot iron.

Until crowning day, nobody knows which breeder has the best tups, so the day marks the beginning of the build-up to the tup sales. Whispered discussions take place over the pen gates: 'A grand packet o' tups,' or ''E'll be yan on 'em.'

Or perhaps the understated: 'Well, 'e 'asnt got one to just set 'em away,' or 'Mebbe not as strang as sometimes.'

What we want is something to get people talking, to spark an interest for when our tups go to the auction. Word spreads like wildfire – they say that there's only one thing worse than being talked about, and that's not being talked about.

We present our shearlings to the crowners in their natural state, just as they come out of the field. There is no beautification at this stage, because it is still a month to the tup sales. Of course, we want them to stand out as good tups, but the last thing we want is to stand out for the wrong reasons. One year, unbeknown to us, one of the children had been playing with the marking stick and had wiped it on the bars of the gate in the pens. Somehow, by the time the crowners and their entourage arrived, the tiny smudge of red marking had migrated from the gate onto two of the tups, which were now sporting red go-faster stripes down their sides. Clive was not impressed.

'Them bloody kids,' he muttered.

It wasn't the first time we'd managed to make a mess of things. A problem with horned tups is that as the horns grow and curl around, they can dig into the sheep's face. It's easily remedied by sawing a sliver of horn away from the inside, nearest the sheep's cheek. The procedure is not painful, and is the lesser of two evils, because an ingrowing horn creates an open sore that attracts flies and, consequently, maggots. It is a two-man job to trim a horn, as the tup has to be held firmly with his head tipped back. Sitting him on his bottom is uncomfortable for him and uncomfortable for us, as he thrashes about and struggles. The concrete floor is hard and unforgiving, but we had a brainwave one year and decided to use one of the many full wool sheets that were sitting in the barn as a cushion for him.

It was only after we'd used the cushion while trimming the horn of our first and best tup that we realized that someone (and it didn't take much working out who, Edith . . .) had graffitied on the sheet with a purple spray marker. Our best tup was now bright purple. We tried to wash it out, but this made it worse, spreading the colour and giving the tup an Edna Everage style rinse.

Determined not to make the same mistake again with the next tup, we went back to the old method of sitting him on the floor in the pen. Unfortunately, sometimes the horn grows at such an angle that the only solution is to remove the whole horn. It is a decision never taken lightly, for looks are everything with the tup shearlings. This one needed his horn completely removed, and after taking it in turns with the cheese wire, we soon had it off. His ear, which had been tucked under his horn, had not seen fresh air for a while and was decidedly stinky.

'Pooh, that's a bit rank,' I said. 'I'll ga an' see if I can find owt for it.'

I went to the medicine fridge while Clive held the tup still and waited. I was soon back with various potions:

'I've got blue antiseptic spray,' I said.

'No,' Clive said, understandably. One purple tup was enough; he didn't want another with a blue ear.

'Green salve?' I said. He raised his eyebrows.

'I want 'em to look like Swaledale tups, not a bloody rainbow family.'

'How's about Stockholm Tar, then?' I asked, holding the tin aloft and waving it.

'If yer want a job doin', then do it yerself,' he muttered, letting the tup loose and going to look for something antiseptic and invisible. I decided I'd leave him to it, and went back to the house.

It wasn't long before Clive was back in the farmhouse, and he was fuming. The tin of Stockholm Tar that I'd left in the hook-over trough had been dislodged; whether by a chicken or an inquisitive tup, who knows? But the result was that we now had a pen full of tups smeared in sticky black tar, as well as one purple one.

'We'el you've certainly got some strikin' tups for us this year, Clive,' commented one of the crowners. 'These fellows will definitely stand out in a crowd.'

10

October

October's arrival brings a nip in the air and the first forebodings of winter. Shadows lengthen as a weak, low-lying sun casts a mellow light across the open moor. The days are shortening; but purple flecks of flowering heather remain on the drier slopes, contrasting beautifully with the overcast skies. It's a good time of year for the amateur photographer, and I always have a camera to hand.

It is a busy time for us. The main focus is still the sheep sales at our local auctions at Hawes and Kirkby Stephen. We sell our older breeding yows (draughts) and surplus gimmer hoggs (shots), and then the tup shearlings. It is only at the tup sales that we dip our hands into our pockets and make a purchase or two. How deep those pockets are is entirely dependent on how we fared at the sheep sales. We need to reinvest the money into new bloodlines, and make good decisions to improve our flock. Informal tup viewings before the sales are an opportunity to get a sneak preview of what will be coming, although you only get an invite to see something good. Anyone whose tups are just middlin' won't be so keen to show them off. These viewings are usually conducted after hours, when the work for the day is done. A driver will be nominated, as the whisky bottle is likely to make an appearance.

Our friend Marshall was an enthusiastic participant in tup viewings. A big, jovial, ruddy-faced fellow with ginger curls, thick glasses and as blind as a bat, he worked for a local landowner and had always been involved in Swaledale sheep, at one time keeping a small flock of his own. Marshall had been married but now lived on his own, but he still had an eye for the ladies. Every Sunday morning he would come to Ravenseat for coffee and cakes, bringing a newspaper with him. He and Clive would go through the lonely hearts column, marking likely candidates with a biro.

'This un' don't sound so bad,' he said, with the paper just inches from his face. 'Fun nights in, that'd do for me.' He grinned, putting a circle around the telephone number.

'Sounds promising,' said Clive. 'Are yer gonna ring t'number an' leave a message?'

Together Clive and Marshall composed a message that they believed would win her over, even if it wasn't strictly true.

'I laugh in the face of danger, and excitement is my middle name,' Marshall suggested.

'That'll do nicely,' said Clive.

Marshall picked up the phone and reeled off the patter.

He wasn't very successful in his quest to find a lady, only on one occasion actually managing to arrange a date with a woman in Darlington. The following Sunday, Clive was very keen to find out how it had gone. Marshall thought for a bit.

'Put it this way,' he said. 'She could stand on t'market square at Kirkby, watch for t'bus an' look at t'church clock at same time.'

'She was cross-eyed, was she?' said Clive. 'Never mind, Marsh; faint heart never won fair maiden.'

When not chasing women, Marshall's passions were growing giant leeks and playing dominoes. But if there was a tup viewing, he was first in line for the outing and spent many a happy evening sitting in semi-darkness, perched on a bale of hay in a shed, fag in one hand and whisky in the other whilst a procession of tups were brought out for inspection.

As more whisky was consumed the atmosphere would get more raucous – comments would be more honest, and perhaps even brutal.

'Hell, looks like 'is horns 'ave just bin thrawn at 'im.'

'He's gotta funny lug, 'im.'

One of the most popular destinations for tup viewings is over in Weardale, where many of our friends and fellow tup breeders farm. The price of an entertaining evening over the border in County Durham is an extremely arduous journey back home, all the passengers with a few too many whiskies on board.

How long the weather holds dictates when the cows and horses are brought in to the farm buildings. Once the fields start getting paddled with mud, then it's time for them to check into their winter accommodation. Winters here are long, especially for the cows, who will not go outside again until May; although living in the barn is no great hardship for them, with silage or haylage on demand and perhaps a bucket or two of sugar beet pellets. A sleeping area is bedded up with straw and the rushes that we'll have mowed off the banks in the summer, and the cows contentedly sift through the rushes looking for occasional tasty morsels.

Before transport meant that straw was more readily available to hill farms, bracken from the moor was mown for the sole purpose of bedding the housed animals up during the winter. This wasn't without its dangers, though, as the bracken could prove poisonous if it was eaten by the animals and had not been sufficiently dried after harvesting.

Every day the cows are moved into a holding pen while we scrape the floor clean. For the horses, winter means overnighting in the stables and spending the day in the pens or at the moor bottom with a hay net. Being of indeterminate, but certainly native, breeding, they grow thick winter coats, and only on the harshest of days do they need rugging up.

Entering the stalls on a frosty morning, the indescribable warm and comforting smell of horses fills the nostrils. Lovers of routine, they will nicker and snort as they wait for their morning rations, occasionally showing impatience by stamping their hooves. Footstamping out of impatience is one thing but when footstamping on a larger scale begins, and they are all doing it, it can mean that the horses have an infestation of lice. Lice can drive a horse almost to insanity, the urge to itch overriding everything else. When they are turned out they hardly take a step without dipping their heads to bite at the feather on their front hooves, or contorting themselves to nibble away at the feather on their back hooves. They can strip all the hair away by rubbing and biting, until only angry, scabby skin can be seen. Then infection may set in. Lice can be found on any part of the horse, but the heels and feather seem to be the most susceptible. Where the lice come from is a mystery: it's a winter ailment, so possibly the straw bedding, or maybe the hay. They are determined little critters, and not easy to exterminate.

One year, after we'd tried and tested many off-the-shelf remedies, it wasn't just the horses that were tearing their hair out. Nothing seemed to work, and I was at my wits' end. I even consulted my antiquarian horsemanship books, but many of the ingredients recommended were no longer available through the normal channels, owing to them being absolutely bloody lethal. Antimony, potassium and arsenic: I don't think so.

Reuben had recently discovered an old half-full glass medicine bottle, perched on a stone ledge in a barn. It had probably been set down by one of our predecessors, put out of harm's way, after being administered to a sickly animal. Whether it worked, we will never know. But the ailments that the magic elixir claimed to cure were comprehensive, to say the least. I scrutinized the yellowing paper label: *Driffield Oils. For sheep, cattle, horses, pigs and all fowl.*

'That's everything, really,' I thought.

For internal and external uses.

'Can't go wrong there, then . . .' I reckoned.

A tablespoonful in half a pint of warm gruel for difficult lambings, scour in calves, foals and sheep. For dropsy, swelled legs, surfeits, impurity of the blood and obstructed perspiration . . . Might need an extra spoonful for this.

Swollen and inflamed navels, colic or gravel, ulcers, fly galls, bites of dogs, wounds, swellings and sagged udders . . .

This last one caught my attention. Clive's, too, but he thought better than to say owt.

'There's nowt it didn't remedy, a cure-all if ever I saw one,' I said to Clive.

There were no ingredients listed on the label, and just about the only problem it didn't claim to solve was a lousy horse. I put the bottle away in our wooden animal medicine chest, preserved for posterity.

A few days later Buffy and Albie, the scrap lads, drove into the yard. Always on the lookout for anything that they can sell for a few quid, they take away anything from broken round feeders to bent troughs, rusty gates and old fencing wire. No job is too big for them: they once took away an old barrel muck spreader on the back of their flatbed Transit by cutting it in half with an oxyacetylene torch. Very occasionally they have something on board that is just too good to be melted down and recycled. I bought a really heavy cast-iron circular pig feeder from them one day. I once even spotted them manhandling what appeared to be a very heavy safe down the high street in Kirkby Stephen. I can only assume that the fact it was in broad daylight meant that I hadn't just witnessed a great heist.

There is nothing even slightly effeminate about Buffy, so out of curiosity I asked his sidekick where the nickname came from, thinking it must have something to do with vampires.

'What's wi' this Buffy thing?' I asked, when his mate was out of earshot.

''E's a Big, Ugly, Fat . . .' he said.

'Reet, got it,' I interrupted him before he finished. It was an acronym. There was no denying that he wasn't easy on the eye.

''Ave yer got any 'osses about?' said Albie. Both lads were from Romany families and were brought up around horses. They are a fount of knowledge, and I like to pick their brains. Their cures are not always conventional, more akin to herbalism I guess, from feeding nettles to cleanse the blood of a horse with laminitis (a hoof disease that can cause lameness and even death), to horse soup made from stagnant water. I like the way these two unlikely-looking lads have absorbed all this information from their elders.

I took them into the stalls, where the horses were picking at the hay in the racks, waiting to be turned out for the day.

'Nice, I's liking yer mare,' said Buffy. 'She's proper lousy though.'

I told him the trouble I was having and how I'd tried every potion known to man.

'Worrabout benzo benzo?' he said. ''Ave yer 'ad a go wi' that?'

'Benzo benzo?' I'd never heard of it.

Walking back towards the flatbed Transit, he went into more detail. Apparently I needed a bottle of benzyl benzoate (available at the chemist's shop), pig oil and a tub of flowers of sulphur, both of which I could get from the agricultural supply shop. All I had to do was mix the powdered sulphur into the pig oil to make a paste, and then tip in the benzyl benzoate. It was all very vague, but I decided it was worth a try.

The next time Clive went to Kirkby Stephen, I handed him a shopping list.

'What d'ya want?' he said, studying the scrap of paper.

'Pig oil an' sulphur, you'll just 'ave to ask at the pharmacy for benzo benzo,' I said. 'They'll know what you're on about.'

Off he went, returning a couple of hours later.

'Yer nivver tellt me 'ow much benzo benzo yer wanted,' he complained.

I explained that I didn't know, as I didn't have an exact recipe. 'Enough to coat all o' t'osses lower legs, I guess.'

'Aye, well that's what I thought an' all,' he said. 'But it didn't ga that weeell.'

'What d'ya mean?'

'I went into t'chemist an' it was varry busy, there was owd Thunderbolt an' Sonny in there, an' I was havin' a bit o' craic with 'em an' then t'lady behind counter asked mi worra wanted.'

'Yes . . . ?' I said, unable to see a problem so far.

'I asked 'er for thi' stuff an' she asked what it was for?'

'Mmmm, what did yer say?'

'I tellt 'er it was for itchin' an' scratchin'.'

Apparently the lady behind the counter raised her eyebrows, looked Clive up and down, and then asked how much he needed.

'Aye, well I need a gay bit,' he'd said. 'I'm gonna slather it on 'cos them lice is making all t'hair fall out.'

Wrinkling her nose up and grimacing, the lady had gone into the back to talk to the pharmacist.

'Would you like to come into this side room?' asked the pharmacist.

'No,' said Clive, getting impatient.

'Mr Owen,' she said, taking her glasses off and leaning towards him across the counter. 'Have you been to see the doctor . . . because I certainly think that you should. Pubic lice are notoriously difficult to get rid of and I certainly won't be dispensing litres of benzyl benzoate for self-medication.'

'Eh?' said Clive, who's a little hard of hearing.

'Pubic lice,' she said, a little bit louder this time.

Clive turned his good ear closer. 'Say tha' again,' he said.

'CRABS, Mr Owen, CRABS . . . you need to talk to your doctor.'

'Oh Clive, I had no idea that it was for that!' I said.

Once Clive had cottoned on, he explained the misunderstanding – but not before the shop had emptied of people, all ready to report back to anyone who would listen that there was a nasty outbreak of something very unpleasant at Ravenseat.

Surprisingly, the treatment worked, and the horses have never been troubled with lice since. I pointed out to Clive that I now keep the potion on standby . . . just in case he should ever need it.

In October, I usually close the shepherd's hut down for the winter. Some folks have romantic notions of staying there when there's snow on the ground, but we know snow and ice bring problems. Even rain brings its own troubles for visitors here, not because there's ever any danger of being washed away, but because we are almost surrounded by water. We have had visitors trapped here, unable to take their cars through the ford as the river has risen so much that it would be folly to attempt the crossing. We can tackle the packhorse bridge in the Land Rover or pickup, as they are not as low-slung as a car, where the oil sump and the exhaust are vulnerable at the apex of the bridge. Even in our vehicles you have to line up with the dead centre of the track, and there is a point when you can only see the sky as you go upwards at an acute angle. You must grip the steering wheel, hold it straight, and then the bonnet will dip and you nosedive down the other side. There may be the odd scraping noise, but we don't worry. Our vehicles are workhorses, and have the battle scars to prove it.

During one rainy spell, we had a run of folk trapped here because of the fluctuating water levels in the river. The land was so saturated that within minutes of the rain starting to fall, the water began to rise. Clive and I watched the levels closely and then had to make an informed decision about whether the car could cross the ford. What sort of car was it: was it light or heavy? Who was driving it: were they confident, or were they going to panic when they got to the middle and stall it? And most importantly: what sort of company were they going to be if they were stuck at Ravenseat with us until the water dropped?

On one occasion we lent our pickup to visitors who needed to be at work later in the day. I was quietly mortified that these professional people with high-flying jobs were going to sit among all the dog hairs, junk and general farming paraphernalia.

If the ford is crossable, then sometimes it is easier to let Clive drive the visitor's car, rather than risk someone having an attack of nerves halfway over and coming to a standstill with the car filling up with water. We pride ourselves on making the right

call, knowing when it is safe to cross and when it is too dangerous. We reckon we've never got it wrong.

At least, not until Alec (who should have known better) managed to get trapped on the wrong side of the beck. Alec himself
was busy doing some joinery in the shed – something that is
more in his line than Clive's, as Clive can't bray a nail in straight.

'Thee tek mi van over t'ford,' Alec said to Clive, and carried
on with his precision woodwork.

Clive jumped into Alec's small van, and set off down the yard.
The beck was running full, but not dangerously so. Clive stopped
momentarily before committing and putting her into bottom
gear, then set off. Just as he reached mid-point the engine stopped
dead, with no warning whatsoever.

With the car at a complete standstill, he had no choice but to
abandon ship. He pushed the door open, and the torrent took
his breath away, leaving him gasping as ice-cold water began to
pour into the van. He shut the door sharpish, the force of the
water helping him slam it. Then he decided that he should make
his escape through the passenger door, which was facing downstream, all the time hoping that the van wouldn't be picked up
by the flood and float off towards Keld. He managed to get out
and run for the tractor, leaving Alec's van slowly filling up with
the dirty river water. Fortunately the tractor had been abandoned at the other side of the river. Without calling Alec, who
was oblivious to the whole drama, Clive reversed the tractor,
waded back into the river and attached a tow chain to the front
of the waterlogged van. He dragged it out to the other side,
leaving the van to drain, and went back to face the music.

'Did ta' manage?' Alec asked, without looking up.

'Weeeell, she's at t'other side, Alec,' said a soaked Clive, balan-
cing on one leg as he tipped the filthy water out of his welly.
'But I did 'ave a spot o' bother.'

Alec looked up, pushed up the glasses that were perched on the end of his nose and studied the dripping Clive.

'What's ta been doin', boy?' he said.

Clive broke the news, and they went down to inspect the damage. They decided they would tow it to the garage and let a professional have a look.

'Dun't try an' start it, Alec,' Clive said. 'The engine might still be OK.'

With Clive in the tractor and Alec in his well-washed van, sitting on a plastic feed bag, they slowly made their way to Kirkby Stephen. They hadn't gone very far when Clive felt a bit of resistance on the tow chain. It jerked a few times as Alec tried to start the engine.

It didn't start, and we'll never know whether it was already busted or whether his efforts did the damage. Either way, the mechanics reported back that everything in the engine that could bend had done so, and it was a complete write-off.

So that was the end of Clive volunteering to take other people's motors across the river.

Our two packhorse bridges have withstood the ravages of time and weather for centuries. In the past, people perished crossing rivers in spate, and the little bridges were built to afford safe passage for the pack ponies and drovers and their flocks. The yows have their trods, the paths that they follow as they wind their way back to the farmstead from the moor, and they know well the bridges and crossing spots. To the stray yows, those that do not belong at Ravenseat, the terrain is unfamiliar and can cause problems. Recently a neighbour came to pick up his stray sheep that we had gathered in from the moor. It had rained steadily for the duration of the day, and the sheep had been left in the pens in the farmyard, waiting for him to arrive.

It was dark when we saw his vehicle lights coming down our road. We walked down to the bridge to meet him.

'We're gonna 'ave to run 'em down t'front o' t'ouse an' load 'em at t'other side o' t'bridge. There's ower much water in t'river for thi to be able to get across in yer motor,' Clive said.

He reversed the pickup and trailer up to the bridge, put a couple of wooden hurdles at either side to funnel the sheep into the trailer and then went to get them. It was still pouring with rain, but a couple of the bigger children had braved the weather and were in position, standing in the darkness blocking any likely gaps where a wayward yow could attempt a getaway. Not being on their home turf makes sheep flighty, and Bill struggled to keep them under control as we moved them towards the bridge. It was difficult to see, as the only light we had was from a head torch that I'd pulled on over my woolly hat, and the dim red glow from the rear lights of the stock trailer. Somehow a yow broke back from the flock and darted off to the left of the bridge. Bill set off into the blackness after her, Reuben and Miles appearing out of the gloom.

'Where's she gan?' Reuben asked.

'That way, towards the river,' I shouted, gesturing.

We set off after her, the beam from the head torch eventually picking out the green reflection of her eyes. She was standing stock-still, and so was Bill. She had a stark choice: move towards the noise and clatter of the trailer and the people beside the bridge, or take a leap into the unknown. She chose the latter, and it was the wrong choice. As I was busy telling Bill to back off and give her some space, Reuben and Miles were creeping around in the shadows in an attempt to get between her and the water, but she beat them to it and jumped. It was a huge jump, landing her in the middle of the swollen river. In the semi-darkness she disappeared under the swirl of inky black water.

We stood on the bank, scanning the roaring water for sight of her.

'There she is,' shouted Miles, pointing frantically downstream.

I could make out her head, tipped backwards, her mouth open, gasping for air. Then she was gone.

'Has she drowned?' asked Miles.

'I dunno,' I said, although in all honesty I couldn't see how she could survive.

We walked back to the bridge, where the rest of the stray sheep had by now been loaded into the trailer.

'Yer one down, I'm afraid,' I said to Clive and our neighbour. 'She's gone, and I doubt you'll be seeing her again.'

The next morning, the weather had changed. It was bright, clear and had a fresh but cold feel to it, a typical morning after a storm. The yard was washed clean, the water had subsided, and only the lines of flotsam on the riverbanks told of the previous evening's deluge.

Miles went to feed the chickens before school, returning with the news that there was a very clean sheep grazing along the roadside and he reckoned it was the missing yow.

Somehow she'd managed to get herself onto dry land, and seemed none the worse for her watery encounter. Maybe her saving grace was that she hadn't much wool on – I know from experience that a woolled yow with a saturated fleece is an absolute dead weight, and she'd have struggled to lift herself out of the river.

Once while I was working near Penrith on a shearing gang, I jumped into the river Eamont, a very deep and fast-flowing river, to save a woolled yow that was floundering. I've always been a confident swimmer, not stylish but strong enough, I thought, to rescue a sheep. I hadn't taken much off in the way of clothes due to the gathering of onlookers. I can't recall the circumstances that led to a sheep being in the river, but I remember very clearly treading water while trying to support her head; and how, when we reached the muddy shore, her legs collapsed under the weight of her own wool. She was fine. I hung a few of my wet clothes on a nearby gorse bush and for a while felt quite the hero – until the farmer came along and said, 'Thoo should 'ave let the bugger drown, it's allus been a beggar for escapin'.'

We tag and record every yow and lamb that goes to the moor, and there are always some that never return. Some we write off, assuming they've drowned in a river or bog. But sometimes we're pleasantly surprised to find them when they turn in late, having just strayed off their heaf; for sheep are free spirits and wander off on their own little adventures, only coming back to the safety of the flock when they feel threatened or are tempted by the rustle of a feed bag.

Tonsing the sheep (pulling out the white hairs from their black bits) is a time-consuming but essential beautification process required to make them ready for the sales. We have a couple of tonsing crates, contraptions specially designed for this job: the tups or yows are hoisted up to eye level and restrained with a collar, their heads resting on a pivoted saddle-shaped cushion. We might use the crates, or if the sheep is amenable, we will lay it down on its side while we sit on a small bale of straw. It largely depends on how much work is needed on the sheep. Throughout October I accept that there will be no tweezers in my make-up bag: the sheep will have perfectly shaped brows, but I will not be looking so well groomed. Fed up with sharing, I had a notion that I would stash my good tweezers away and buy cheap stainless-steel pairs from the local chemist's – perfectly adequate for the removal of stray white facial hairs, I thought.

But I was wrong. I had a small group of unhappy men in the yard, complaining that the new tweezers were not grippy enough.

'Nay to hell, they're as slape as snot,' said Clive. 'I cannae git them hairs for love na' money.'

Alec and Steve nodded in unison.

'They're useless,' Clive continued, even after roughening up the edges on the stones on the building corner. The substandard stainless-steel ones were soon lost in the straw covering the floor of the barn, or stuffed into a straw bale for temporary safe-keeping and then forgotten about – and the men were back demanding my prized Tweezermans.

As Clive's birthday is in October it made perfect sense to me to buy him his very own tweezers, good ones in super-bright colours so he could easily spot them in the straw. Now he looks like a regular man about town, with the tweezers living in his top waistcoat pocket at all times. He is not alone: recently Reuben came home from the auction saying that he needed to find the metal detector, as Eric, a farmer further down the dale, needed to borrow it.

'Whatever for?' I asked.

'Dunno, 'e wouldn't tell mi,' said Reuben. 'Just said 'e'd lost summat.'

Turned out it was his tweezers, his favourite pair. Once upon a time it was a secret that the Swaledales were tidied up before sale time, with the odd stray white hair removed with the fingernails. It is now deemed normal to define the black from the white. As well as tonsing, we trim the bellies to give the sheep extra height and use peat to colour the fleece. The peat is gathered from the haggs at the moor, dampened and mounded into a ball. It's then dabbed onto the fleece, applied with the lightest of touches; it then dries to a subtle shade of heather grey, emphasizing the whiteness on their legs and faces. It is not about faking; more about enhancement.

When all of our surplus breeding females, young and old, have been sold, it's time for the tup sales. For the whole of October we keep a close eye on the young tups (shearlings), needing them to remain quiet and settled. It's not easy, because this is the time of year when the sap is rising and they are desperate to escape and find some wanton yows. We keep the tups inside, away from the yows, but they can vent their frustration on one another. Full-on fights can happen, the tups going literally head to head, sometimes leading to disfigurement – which scuppers any plans to get them to auction looking their best. Even if the damage is just cosmetic, it will ruin any chance of them going into the show ring, and will put off would-be buyers. We try to keep the tups together but if we hear the sickening noise of them smashing headlong into one another, then it's time for a spell of solitary confinement. The problem then is boredom: a bored tup can rub all the hair off his face in a single afternoon.

Eclipse was one of our better tups. ''E's a bit special, this 'un,' said Clive on many occasions in the run-up to the sales.

But Eclipse was on a mission to self-destruct.

First he wanted to fight with all and sundry, so he ended up in a stable, all on his lonesome. Then he decided to rub on everything: the bale of hay, the water bucket, the door frame. His blood was up, his mind was racing and what he needed was a distraction.

Clive made him a punchbag out of an empty feed bag stuffed with hay, suspended from a beam. Tufts of hay stuck out of the bag here and there, and Eclipse would alternate between nibbling and wrestling with it. It did the trick, with the distraction preserving his good looks so that he went to the sale and made £11,000.

Every year the pens of the local auction marts come alive with farmers and their families showing off the best of their breeding and on the lookout for new tups to introduce to their flocks. Farmers will often take home rather more than they bargained for. We call it Tup Sale-itis, the coughs and colds that are spread so easily through a throng of people in a confined space. Many people are laid low the week after tup sales.

The sales are held over a few days, old tups and tup lambs on the first day, then shearlings. After the weeks of beautification, it is a relief when sale day finally arrives. It begins early with a show. Two judges are chosen by the Swaledale Sheep Breeders Association, and their task is to rank the competitors' sheep. Rosettes are awarded down to sixth place, and are much coveted because fastening one to the rails of your pen brings potential buyers your way. The start of the sale is announced by a bell, and the sale ring begins to fill with buyers and spectators. A ballot drawn prior to the sale decides the order in the ring: nobody wants to be first, and nobody wants to be last. Somewhere in the middle is perfect, with the sale warming up but before the buyers are fully committed and losing interest. For vendors, the wait is spent sitting on a hay bale in a well-bedded pen, talking through the tups' pedigree with would-be buyers and getting through lots of cups of coffee from the auction canteen. A tannoy system relays the fortunes or misfortunes of those who are in the ring. When big prices are being reached, silence descends over the pens, everyone listening. The auctioneer's dulcet tones are relayed through the crackling loudspeaker.

'Value the breedin',' he says. 'This is 'im today, I'm not startin' 'im under ten thousand, so come on gentlemen let's be gettin' on . . . Where d'ya wanna be? Someone'll give me five, surely?'

Someone does. 'Thank you very much,' he says, and then he's off. The flap of a catalogue or an almost imperceptible wink means another bid.

'Six, seven, eight . . . ten, is that right?' says the auctioneer, his eyes scanning the room, the drovers in the ring helping by pointing bidders out. Bids come in thick and fast, in ever increasing increments.

'Twelve, fifteen, eighteen.'

By now there is a palpable tension, as the bidders dwindle and the final two remaining begin to waver.

'Don't leave him now, lads, just round him up. Are yer sure you've finished? 'Cos I'm gonna sell 'im.'

Then the gavel falls.

Topping the auction is what so many breeders dream of, but the reality is that only a very small handful of tups reach these magical heights. Many will leave the ring with their tups unsold, the auctioneer's words, 'There'll be another day for 'im,' ringing in their ears.

There are tups for all pockets, whatever your budget, from fifty pounds to fifty thousand. At the higher prices, the tups go to the award-winning pedigree breeders who have to keep bringing in the best-quality breeding lines to stay at the top. Once you've been into the auction office and paid the price, he's yours to take home. One year there was big trouble at the auction when one of the top-money tups went missing. When later in the day his new owner went to pick him up, the tup was nowhere to be seen. Nobody knew where he had gone. There are hundreds of tups in the auction each day, so there was a lot of searching to be done. Later that evening, when everyone had gone home, there was one unclaimed tup left in the pens. Tied into his fleece on the middle of his back was the little circular label with his lot number: 616. But read the other way up, it was 919. The tup in the pen had been bought earlier for a few hundred pounds, but an honest mix-up had sent the buyer home with one worth thousands. The issue was soon resolved, and the tups reunited with their rightful owners.

Every year there are dramas, whether it's an established breeder taking bad prices, or a little-known breeder winning the show. There are no guarantees, and we all have good years and bad ones. We've learned to expect the unexpected.

One year, our friends Colin and Anne spent weeks in the run-up to the tup sales getting excited about their prospects, feeling confident that they had a good packet of tups. They set off to auction in good spirits, and after putting the tups into the pen they went to the canteen for bacon sandwiches and coffee. People were already milling around, so it wasn't long before they headed back to their pen to talk to prospective buyers. Leaning over the gate, supping their coffee, they noticed that one of their tups was asleep at the back of the pen and was not exactly showing himself off to the viewers.

'Anne, ga an' git that tup up,' said Colin.

Anne dutifully put down her paper coffee cup and went into the pen, the tups moving to the other side while the sleeping one didn't stir. It is not unusual for a sheep to sleep quite deeply: we've had them asleep in the fields, and walked right up to them before they woke. Unfortunately, though, this time the sleep was of a permanent nature. Hurrying back to Colin, Anne beckoned him.

'Colin,' she muttered, sotto voce. 'He's dead.'

'What yer on aboot? Are you sure?'

'I knaw a dead tup when I see one,' she replied.

Tipping the remainder of his coffee onto the concrete, Colin opened the pen gate and walked purposefully over to the tup with the sole intention of proving Anne wrong.

A casual nudge in the ribs with the front of his brogue got no response. He gave a firmer kick: still nothing. Anne looked on, hands on hips. Sidling up to her, he said, 'Anne . . . Anne, 'e's as dead as an 'ammer.'

It was time to decide what to do. Their friend Dave was summoned.

'I need yer to 'elp us out 'ere, Dave,' said Colin quietly. 'There's a whole load o' folk around and I need to get rid of a body without anyone seeing.'

'Yer tup's fizzled out?' said Dave, astounded.

'Shhhhhhhh,' murmured Anne.

It was not a great selling point, having a dead tup in the pen, cause of death unknown. They're supposed to be fit, full of fire and ready for action, not belly up.

The auction was buzzing with folk, and so far nobody had realized that the tup lying down at the back of Colin and Anne's pen, his legs folded under him, his chin resting on the straw on the floor, his eyes closed and a peaceful expression on his face, was now grazing heavenly pastures.

Anne hurried off to bring the Land Rover and trailer onto the loading docks while Colin hatched a plan.

'This is what we're gonna do,' he said.

'Are yer sure 'e's dead?' said Dave, looking through the pen bars.

'We've been through this,' said Colin. ''E's definitely dead. We're gonna get hod of a horn each, Anne'll keep his back end up an' we're gonna walk 'im out of the auction like nowt's wrong. If anyone asks where he's going then we'll say that he's going for his picture takin'.'

By now Anne had returned, and the plan was put into action.

'Jeez, he's bloody 'eavy,' said Anne.

'C'mon, lift, woman,' said Colin.

They set off down the alley, pushing through the sea of people. Nobody batted an eyelid. In fact, the tup got one or two compliments en route to the trailer.

'Not a bad sort of a tup you've got there, Colin,' said one. Nobody noticed the veins bulging in Colin's forearm as he held the tup's head aloft: he was lifting a dead weight in every sense of the word.

'Mornin', Anne, good sort you've got there,' said another.

'I was just on mi way to thi' pen,' said yet another.

'We'll be back in a minute,' shouted a red-faced Anne. 'Just takin' im for 'is photo for t'flock book.'

They exchanged pleasantries for the whole of the short journey, and then unceremoniously bundled the tup into the trailer. Slamming the back door, they went back round to the pens, and nothing more was said.

'Well done, Dave,' said Colin. 'I'll buy thi a pint later on.'

The rest of the day went reasonably well, trade was brisk and they sold the rest of their tups. Afterwards, Colin, a man of his word, headed for the pub in search of his pallbearing friend Dave. The post-tup-sale drinking session went on late into the night. Finally, when time was called, they stumbled out of the pub door, a little worse for wear. Anne, the designated driver, was more clear-headed, and guided them along the dimly lit street to where the Land Rover was parked. Fumbling for the keys, she heard a faint thud.

'Colin, did yer 'ear that?' she said, standing stock-still.

'What yer on about?' said Colin, between hiccups.

'I'm tellin yer, there's summat in t'trailer.'

'It's Dave, he's gone for a pee around t'back.'

'No, Colin, I'm tellin' yer – there's summat in t'trailer.'

To appease her, Colin wobbled his way along the pavement and peered through the vents in the side of the trailer.

'There's a flamin' tup in 'ere, Anne, fetch a light.'

''Ow many bloody brandies 'ave you 'ad, Colin?'

'I'm tellin' yer, there's a tup in 'ere,' he reiterated.

'Aye, yer reet, Colin, a dead 'un.'

Going through her pockets, she reached for her phone and shone its torch into the trailer. Looking back at her was a thoroughly bad-tempered tup.

'Christ Almighty, 'e's come back to life!' she exclaimed.

'Well, I've seen everything now,' said Colin, shaking his head.

The next morning, in the sober light of day, they investigated the contents of the trailer. Sure enough, there stood the tup, as large as life and seemingly as fit as a fiddle. When Colin let the trailer door down, he bounded out and up across the field. They had no idea what had happened the previous day: he'd certainly been unconscious, and his breathing imperceptible. It was only a couple of days later that they found him lying flat-out in the field in exactly the same way, but sadly this time there was to be no resurrection.

Where there's livestock there's dead stock, but usually there's no coming back from the latter. You do occasionally come across mysteries, and we will never know what happened to that tup. But it was certainly better that he died at home than with a new owner.

11

November

November for us is all about tupping time: turning the tups out, and getting the yows in lamb. We aim to get the tups out by 5 November, for lambs on 1 April. Bonfire night for April Fool's Day. As a general rule, a hundred yows is the most that can be put to one tup. Any more than that, and you're asking for trouble – an exhausted tup may overlook yows coming into season. We prefer to err on the side of caution, and have only fifty or sixty yows to a tup at any one time.

It isn't just a case of putting a tup in the field with the yows and letting him do his work; it is vitally important that we monitor what is going on. Only one tup joins each small flock of yows because we breed pedigree, and we need to know which tup has sired which lambs. The yows are tailed, which means the wool from the sides of the tail are clipped in order to allow the tup better access. Then they are carefully selected, to make sure there's no inbreeding and that the tup looks to be a good match for the yow. Darker yows are put to lighter tups and vice versa, in the hope that any failings the yow has can be rectified by matching her with a tup with the right credentials.

Splitting the yows into smaller flocks and keeping them separate for tupping time brings its own problems: there are so many individual fields to find, because the sheep still need enough grass to sustain them. But if you put them in a field that's too big, the tup may fail to sniff out a yow in season, and miss mating with her.

In order to tell which yows have been tupped we daub ruddle onto the tup's brisket; when mating takes place this is transferred onto the yow's rump, leaving a mark that is plain to see. Powdered ruddle in bright colours is mixed with oil to a thick enough consistency to stick to the wooden paddle that we use to apply it. Taking the ruddle, together with a scoop of high-energy sheep cake, we visit every tup every day, giving him a bite of the feed to keep his spirits up (and his pecker too for that matter). It's also an opportunity to gather up the yows to him, to make sure that nobody in season gets missed. For some tups, a pan full of food is irresistible – whereas for others, food is the last thing on their mind . . .

Thunder was one of our older tups, bred by Ron Metcalfe, a renowned breeder of Swaledales and great friend of ours who is now sadly no longer with us. We didn't keep Thunder just for sentimental reasons: he was a good-getting tup. We had to use him quite carefully as we had kept many of his offspring as breeding yows, but every year we'd have a few yows for him. If I had to describe Ron, I'd say that he was bold, no-nonsense, out-spoken and quite mercilessly honest in his critique of sheep breeding, and perhaps in his later years, even more so. His tup Thunder seemed to inherit some of his character as he aged, becoming gradually more cantankerous and set in his ways. When I looked into that tup's eyes, he returned my look with a steely, knowing gaze that unsettled me. I never really trusted him – and rightly so, as it turned out.

Thunder loved his food ratio. It didn't matter what he was doing or where he was: whenever he heard the rattle of the feed scoop his head turned, nostrils flaring and ears switching backwards and forwards. Once he'd locked on to where the food was he'd set off at full speed, never taking his eyes off the prize. This was fine, good even, as there was never any need to catch hold of him in order to apply the ruddle. He'd eat away, occasionally looking up while a few crumbs of food dropped from the corners of his almost toothless mouth. I'd daub the coloured rud on whilst he guzzled; Kate would gather up his small harem of yows, and then would retreat to the relative safety of the quad bike. Dogs are quick learners and Kate knew all about Thunder, having been chased out of the field by him once. She'd saved herself that day by slinking out under the gate with her tail clamped between her legs.

It was a particularly beautiful autumn day. There had been a sharp overnight frost; a coolness remained in the air, and an early-morning mist was hanging over the valley bottoms. Clive was foddering the cows in the barn and I set off with Miles and Kate to rud a couple of tups. We took the quad bike through the Beck Stack and rudded our first tup, Battler. He was awkward, only having been out with the yows for a few days, so that the novelty of having lots of lady friends hadn't worn off. Between us we managed to corner him and the yows at the top of the field and after slinging a bit of feed in his direction and fending off a few greedy yows he finally, cautiously, put his nose in the feed scoop, and we got a couple of blats (splats) of rud on.

'I wanna ga yam an' ave mi toast,' Miles said. 'I 'aven't 'ad mi breakfast.'

'Nivver worry,' I said. 'This won't tek lang, Thunder won't take any temptin', he's good to do.'

Off we went, leaving Battler sniffing about amongst his yows. Every so often he looked up, stretching his neck, his lips furled back. Five or six yows circled him, captivated and competing for his attentions.

Through the gate and into the Close Hills we went, parking the bike just below the abandoned farmhouse. Steam rose from the frosted grass as it was warmed by shafts of bright sunlight. Climbing from the bike, I paused and enjoyed the scene: the haziness had lifted, and I could see the narrow road winding out from Ravenseat and the muted greens and browns of the moors in the distance. This, surely, was a good photo opportunity. Grabbing my camera, I took a picture, then looked across to Miles, who, having temporarily forgotten his pangs of hunger, was poking about in a clump of seaves alongside a broken-down wall. I never forget how lucky I am that this beautiful spot is my workplace.

I stopped my daydreaming and snapped back to reality and the job in hand, swapping the camera for the ruddle pot and filling the scoop with feed out of the half-filled bag that was on the front of the bike. Kate was getting impatient, so I sent her along the wall to bring the few yows that were in sight. I rattled the feed scoop and whistled for the sheep, waiting patiently, keeping tabs on Kate, who was watching the yows as they made their way towards me. Then, in a split second, the most fleeting of moments: out of the corner of my eye I caught a glimpse of wool. I instinctively turned to the side and was dealt the heaviest of blows, one that took my legs clean out from under me. I saw the sky as my body twisted and then fell awkwardly onto the damp grass. Laid out on my back, quite winded, I gasped and looked down towards my wellies to where Thunder stood, his jaw pulled in towards his chest. He looked down at me, then began to back up, his head down. I knew what was coming.

'Blaargh, yer rotten beggar,' I shouted, kicking out at him with my wellies.

He cocked his head to one side, considered having another go and then thought better of it and ambled off to where the upturned rud pot, stick and scoop were lying. He began to hoover up the particles of scattered feed spread across the grass and in the puddle of yellow ruddle that was oozing from the pot. Clambering to my feet, I turned to see that Miles and Kate had scarpered back to the bike.

'Are yer alright, Mam?' Miles shouted.

'I'm fine,' I replied, truthfully. It'd knocked the wind out of my sails but I was more angry with myself than Thunder; I should not have been so complacent with him. I'd always known that he spelled trouble, but that was the first time ever, in twenty years of shepherding sheep, that I'd been taken out with such force by a tup. I thanked my lucky stars that I hadn't brought baby Annas along in the backpack, brushed myself down and went to retrieve the rud pot, stick and feed. Thunder carried on eating the spilt food and I never bothered trying to get any colour on him, as he had daubed himself up without my help. When he paused in his eating, he looked towards me and repeatedly stuck out his coarse black tongue in an attempt to remove the sticky yellow rud that coated his snout. I gave him a hard stare and decided that his card was marked.

Back home Miles couldn't wait to announce to Clive the exciting events of the morning, how I'd been airborne and stomped into the ground when Thunder launched his full-frontal assault. There was plenty of exaggeration, but it all served my purpose: getting Clive to agree that Thunder was going to the auction mart.

'Aye, he's just gone a step too far this time,' he conceded. 'Are you alreet?'

I repeated that I was fine. 'I tell yer what, though. I might 'ave a tup's head tattooed on mi arse, but I reckon I'm gonna 'ave a bruise that's the shape of a tup's head imprinted on mi side now.'

Clive was true to his word, and Thunder went to the auction. He left behind his legacy, the lambs that he sired in those first few weeks of tupping time. Typically, to our annoyance, the best lambs we bred that year were all fathered by him.

Sheep aren't really classed as dangerous animals, but they can inflict some damage if you are in the wrong place at the wrong time. Clive once had a black eye inflicted by a tup. Raven, too, has been flattened when a gimmer lamb did the dreaded three-bounce manoeuvre. This move is mainly confined to lambs or younger sheep. It usually happens when one is cornered, and thinks the only way of escape is to run full tilt, head down, at whatever is blocking the way, whether it's a gate, a wall or a person. The run-up ends with two small straight-legged springs, and then a final enormous one that propels the sheep forwards and upwards. If you are the target, it strikes at about chest height. We see this move a lot at the auction mart, often when the sheep is coming down a narrow alley with nowhere to turn. In Raven's case it happened in the sheep pens as she closed a gate. She was OK. But a sheep's head is a very solid thing to come up against. Our friend Jimmy once spent a week confined to bed after being knocked unconscious by a yow in similar circumstances.

Although we have stock tups, reliable chaps who we have used in previous years, we also introduce our new tups, the ones we've bought at the sales. Some will have been handled a lot, preened and possibly shown. These should, we hope, be good to catch in the field. Others are so embroiled with the exciting task in hand that only cunning and stealth gets them rudded – sneaking close and then grabbing a horn and hanging on for dear life, with your heels dug into the ground. The tups usually run with the yows for about six weeks and some become better behaved as the weeks wear on, or perhaps they are simply hungrier for a bite of feed as the strain of keeping the yows happy takes its toll.

One wet and horrible afternoon I had a radio recording to do on moudie (mole) catching and shepherding for *Woman's Hour* on Radio 4. We'd had our fair share of trouble with the inter- view as it was conducted outside while I was on the job. The rain meant that the waterproof cover shielding the microphone muffled the dialogue, but not enough to cancel out the sound of heavy breathing from my reporter friend. He was a sturdy chap and not really dressed for the conditions, his leather-soled city shoes not providing much grip as we went up the slopes, which were black with mud from where the moudies had put up their hills. I talked away as we went along, demonstrating the art of moudie catching and explaining why we waged war against the grey velvet rascals.

'It's not personal,' I said. 'Moudie tunnels are good for t'drain- age, 'specially in t'lower field. We catch 'em 'cos of t'ills that they put up.' I think I sounded like a real country yokel, but that's probably what they wanted.

In a dry summer we make hay, and any soil from the molehills that is picked up with the grass dries out in the hay loft and either drops out in there, or in the fields when the hay is fod- dered. The problem is that in a wet summer we have to make round haylage and silage bales, and the grass is preserved by fer- mentation. Haylage is the better, drier stuff that is more palatable for the sheep and horses. The cows will happily munch through the wetter, more vinegary silage whereas the pickier horses and sheep would almost certainly turn up their noses. Everything inside the bales warms up and any bacteria within the soil con- tamination multiplies, causing listeriosis and sometimes botulism amongst the sheep, cows and horses.

I explained all this complicated stuff.

'We might as well ruddle Gem, he's just 'ere,' I said to the reporter as we were now near the Hill Top field and the rud pot was on the bike. He didn't have a clue what I was talking about. I explained as quickly as I could what I was about to do.

'Gem is a good boy,' I said. 'No need for a dog, he'll come runnin' for 'is food.'

The journalist nodded and I whistled for the sheep while I got my equipment assembled. I had the pot, stick, bag of feed but no scoop. *Never worry*, I thought, improvisation was needed. I decided to tip the moudie traps out of the blue bucket that I carried them in, fill it with feed, and then feed Gem out of that.

My companion followed me as I strode through the gate and onto the hill end. In front of us was a commanding view of Upper Swaledale, partially shrouded by low cloud. A squally shower of rain blew through, as I impatiently waited for Gem and his small flock of yows to climb out from the bottom of the field, talking into the microphone to set the scene for the listeners to the programme.

'We're standing at the very head of Swaledale,' I said. 'Picture the scene: heather moorland, blowaway grass, *Wuthering Heights*, shepherdess and sheep in perfect harmony, a timeless scene.' I was laying it on thick now.

Gem arrived, puffing and panting almost as much as the reporter had been on the climb. The reporter held the mic in the air, towards the tup, for sound effects.

I held the bucket casually in one hand, armed with the ruddle pot and stick in the other. Gem looked suspiciously at the reporter and then stuck his head in the bucket of feed. I morphed into David Attenborough, whispering into the mic.

'And here we are in Swaledale, it's mating time for the sheep, here we have a tup . . .'

Before I could say any more Gem put his head up, catching

me unawares, and got his horns wedged through the bucket handle with his head inside. That was the end of the idyllic scene. He panicked and went straight into reverse, ripping the bucket from my hand, then letting out the deepest, manliest 'baaaaaa' imaginable, holding a note that echoed from the depths of the bucket. The acoustics were perfect. He turned to the right, then left, although he could see nothing other than the bottom of the blue bucket. Then he lifted his head upwards, the feed spilling out onto the ground, before setting off in, literally, blind panic.

The reporter looked startled. The sight of a tup careering around the field with a bucket on his head was comical, but I was thinking that it made me look very amateurish.

The yows had cleared off, not impressed by Gem's antics with the bucket. I was deciding on my next course of action, which was going to be a trip home for a sheepdog, when thankfully the bucket handle broke off on one side. Gem gave a massive shake of his head, and the bucket and tup parted company.

Luckily, the tup-rudding episode was consigned to the cutting-room floor before the programme was broadcast. There's an old show-business saying about never working with children or animals, but if you have a family and live on a farm, then that's impossible. There's certainly nothing predictable about either, and they will both show you up from time to time.

The yows cycle every seventeen days, and we have a system of changing the ruddle colour every eight days or so, beginning with a lighter colour such as yellow, then blue, then finishing with red. This allows us to see that yows that were rudded in the first few days are in lamb and have not come into season again. If a yow is rudded twice in different colours we can tell that although she has been mated, she did not become pregnant in the first instance. If it happens a lot within one flock, we can assume the tup is infertile. The colour coding system also gives us a due date: yows with yellow behinds will lamb first, then blue and red. First weekers, second weekers and third weekers. There is never any doubt as to what colour rud we are using on the tups in any particular week. At the end of the first week the children are yellow, like Bart Simpson, second week they look like the Smurfs, bright blue, and by the end of the third week they are a little green around the gills.

There is an alternative to rudding the tups every day, and that is a tup harness with a square, soft crayon that fits onto the front. We have never had much success with these, as they chafe the tup and often in our cold climate the crayon hardens and does not leave enough of a mark to be identifiable some five months down the line. And as Clive says, 'It puts the tup off, they don't perform as weel, an' yer wouldn't yer sel' wi' a brick teed to thi chest.'

But he also says, 'It in't t'rud that gits lambs.'

This comment is usually reserved for when I've been a little too liberal with the application of rud, resulting in the whole yow turning yellow, blue or red.

The Swaledale Sheep Breeders Association forbids the use of artificial insemination, preferring things to be done the old-fashioned way. This is, in the main, to avoid extensive use of a single tup, risking inbreeding. Everything is done as nature intended, for Swaledales are a native breed and although fashions change, even amongst sheep, certain traits must remain the same. A Swaledale needs to be able to thrive and be productive in the harshest of environments.

When we are happy that the yows are in lamb we begin to turn them back onto their heafs at the moor, but before we do so we need to record and smit each yow. Smitting is the marking of each yow with a coloured paint that identifies who the yow is in lamb to. We keep a diary, a little black book full of important dates, when tups were loosed, when cows were bulled and such-like. It's in this book that all the smits are recorded. Clive has never been good at telling his left from his right, so I draw a little diagram of a yow and mark it with the appropriate smit, which can be anywhere on its body and in any colour, and underneath write the name of the tup. I sometimes get artistic and draw the same sheep with smits on the wall of the sheep hospital building at lambing time, so that everyone will recognize them and know them automatically. I read the yow's ear tag with the electronic scanner and record who she's in lamb to – a belt and braces approach, as there are times when you search through the fleece of a yow and can't find the smit.

The yows are always happy to leave the confines of the in-bye fields and return to the open moor, where they will stay until they lamb. A small number of yows may 'break', losing the lamb in the early stages of pregnancy, and for this reason a tup will be turned to the moor with them to 'jack up', catching any that come into season at a later date.

Sometimes, depending on how many tups we have, we may even save a 'fresh' tup just for this job. He is raring to go, and no fertile yow escapes his attentions. Every day he is fed and rudded and, with winter fast approaching, the yows are also given a small amount of feed, which brings them all together for the tup.

Keldside Image was a reliable and smittle (fertile) tup; we thought a bit of him. He was tall with a big head, sawed horns, a broad muzzle and a commanding presence. He was a bit special and he knew it, so we decided one year that he would go back to the moor with the yows. He'd only been with them for a week when one morning we went up there and he was nowhere to be seen. We assumed he was off gallivanting, and was probably holed up with a yow somewhere. We didn't worry.

'He'll be back tomorrow,' Clive said. 'He'll be under t'wind in a ghyll somewhere, he'll nivver 'ave 'eard us.'

The next day there was still no sign of him, and the children were keen to go on a ride on the bike to see if we could spot him.

'We're gonna play I-spy,' I said to Violet. 'Summat beginning with T.'

'Tree?' said Violet.

'Well there isn't any o' them,' I said. 'Nope, tuppy tup.'

We caught up with a few stragglers, yows that weren't as keen on feed, and had broken away from the main flock. But Keldside Image never appeared. When I got back to the yard, Clive was surprised.

'I wouldn't 'ave 'ad 'im down as goin' walkabout,' he said.

'Mebbe there's an 'ole in t'fence,' said Reuben, the optimist.

'Mebbe 'e's deeead,' said Miles flatly. He's the pessimist.

Raven decided that our search of the moor had not been thorough enough and that we should change our mode of transport in order to get a better view.

'We need to go on t'orses,' she said.

This was inspired thinking, and we were spurred on by Clive upping the game by putting a bounty on finding Keldside Image.

'Bribery gets yer everywhere,' he said, announcing over dinner that anyone finding the missing tup would be rewarded with a crisp tenner.

We tacked up Meg and Josie, attaching panniers to the D-rings on Josie's saddle to carry a pair of old-fashioned, heavy binoculars. Being on horseback instantly gave us a better vantage point. It's a peaceful mode of transport: apart from the occasional snort from the horses all was quiet, the horses treading softly along the worn sheep trods. Although we had scoured the moor with the bike, there are so many ravines and screes interspersed amongst the peat haggs that there was still plenty of uncharted territory to cover. The sun was shining and our cold fingers held the reins loosely, letting the horses take care of the route as we scanned the hidden gullies and hillocks alongside the sinuous beck that wound its way down the valley bottom.

We could see for miles. A pair of ravens circled overhead, prompting me to wonder whether Keldside Image was now nothing more than a corpse, his bones being picked over by the birds. We made a long steady climb out towards the boundary fence, stopping momentarily to let the horses catch their breath. The fine weather meant the yows had scattered after their morning feed, and were dotted here and there, heads down, taking no notice of us. Our search was, sadly, fruitless. It was not a waste of time, though: any time spent studying your sheep and the lie of the land is always a bonus.

The only glimmer of hope was that we found a large hole in the fence, and on close inspection, we could see wisps of wool caught up in the wire.

'There's feetings an' all,' said Raven, still in the saddle but leaning forward and pointing down to footprints in the exposed peat at the fence side.

'Aye, someone found this 'ole afore we did,' I said as I patched up the hole with baler twine and pulled off the tendrils of wool.

We told Clive when we got back, and put word out that we had a tup a'wantin. There were reports of sightings of Keldside Image in all manner of random places: the adjoining moor, out on the common or in any number of fields between here and Keld. But each sighting turned out to be a false alarm.

It was almost Christmas when we finally found him and, for us, the parable of the lost sheep didn't have a happy ending. We had hoped upon hope that he had gone walkabout, but one morning when the moor was greyed over with a light covering of snow, I deviated from my usual path. I was looking for some solid clean ground to put out a line of sugar beet pellets without them turning to mush. Seeing a sizeable clump of seaves surrounding a small depression in the ground I swerved away, knowing that this was the sign of a bog. By sheer chance I caught sight of a clump of wool floating among the greenery, for it was only in these very wet places that the snow didn't stick. In the middle of the bog was Keldside Image. He must have walked straight into the bog and sunk. I imagined that beneath the surface of the mire he was suspended in a standing position, only his back and the top of his head visible above the green moss and water. I got close enough to reach out and touch him: he was just inches away from safe ground. I hope his end was quick and painless, but I felt very frustrated at having been so close and yet not having seen him in time to rescue him.

In circumstances like these, the watery grave would have to be his resting place for eternity. Getting him out would have been impossible, so there he stayed. I don't know whether it would be classed as an illegal on-farm burial, because no burial actually took place; his mortal remains just slowly slipped below the surface. There are many, many places like this on the moor: wet shops, bogs, whatever you want to call them. It is impossible to know where they all are: occasionally they disappear when a spring dries up or a watercourse moves, but new ones appear all the time.

We are fortunate in many ways that bogs are the only real hidden danger that lie in the heather. Once you move from our moor towards Tan Hill or Keld, you can get a sharp and unpleasant reminder of the now defunct coal and lead-mining industries of the last century, as our friend Alec did. The scars above the ground are still visible, but slowly over time the landscape consumes them, claiming back the buildings and spoil heaps that the heavy industry left in its wake.

As the outward signs gradually fall into decay and evidence of the sweat and toil of the past is forgotten, the subterranean world remains surprisingly intact. When the mines ceased production the entrances were blocked up, the open shafts capped and any obvious dangers made safe. Sometimes the cap that covered the gaping hole was as simple as a few wooden beams, covered over with soil and rocks. Over the years these rotted, the soil moved with subsidence, and eventually the shaft opened up.

Alec had been enlisted to help gather the sheep in from the Arkengarthdale moor. At that time he was still the landlord of the Tan Hill Inn, the pub at the very top of Arkengarthdale, which meant he was very familiar with the territory. It was his home turf, and he was on friendly terms with many of the local farmers who frequented the pub. Always having a legion of well-trained sheepdogs at his disposal, he was often called in to help out on bigger gathers, when all the farmers would work together to bring in the sheep from the moor. On this particular gather the day had been a long one with a lot of ground covered, and reasonably successful in that they had a very large flock of sheep by the time the pens were in sight. There were farmers on bikes, shepherds on foot, sheep bleating and dogs barking.

With all gathers, there are certain places that pose a danger when it comes to sheep escaping. Alec was ahead of the game, knowing exactly where he needed to be to prevent the sheep making a break for freedom. Once in position, he turned any wayward sheep that came his way, with his dog, Mack, driving them back towards the main flock before they could break loose. Often working at a distance, Mack would occasionally be out of Alec's sight and responding just to his whistles. Time and again Mack would turn the sheep and head them off in the right direction. As the last sheep headed over the hill, Alec whistled for Mack and called his name. Mack did not appear, but Alec wasn't worried, assuming that he had joined forces with the other sheepdogs at the back of the main flock. He was slightly annoyed that his dog was not listening, and set off to join the other men, who were now driving the sheep into the pens.

'Did my dog com' wi' you lads?' he asked.

'What? Mack? Nivver seen 'im, Al,' was the reply.

While the men sorted the flock, debating the day's gather, Alec went back to where he'd last seen Mack and let out a piercing whistle that carried on the wind. He stood watching and waiting, willing there to be some sign of the dog. Occasionally there'd be a stirring amongst the heather and the sounds of movement, and Alec's spirits would lift; then a startled grouse would take to the air, dashing his hopes again. It was all very perplexing, and all he could think was that Mack had perhaps taken it upon himself to go home alone. This wasn't beyond the realms of possibility, as Tan Hill was not far away as the crow flies.

Back in the busy pub, Alec's wife Maggie, the staff and customers were all adamant that there'd been no sightings of Mack in the vicinity, on the road or footpaths. It was out of character for Mack – he was a dog of good temperament, and very biddable. Alec had to assume that something had happened to him. He decided to go back to where he'd last seen Mack for one last look. There is nothing worse than a sheepdog a'wantin at the moor, as you know that the dog will chase sheep.

Taking his time, Alec walked slowly through the heather, following the same direction that he reckoned Mack had gone in when pursuing the breakaway sheep, until he reached a patch of bare ground where he noticed a small circular clump of bracken growing around an indentation. Poking his stick amongst the greenery, he was surprised to discover a deep hole. Getting down onto his knees and parting the stems and leaves, he peered down into the darkness. He was unable to see the bottom, but he could hear the sound of running water a long way below. Now laid out on the grass, leaning over as far as he dared, he shouted, 'Mack! Mack!' His voice echoed. 'Mack, is ta down there?'

He was sure that somewhere below him he could hear Mack's wet tail splashing in the water as he feebly wagged it at the reassuring sound of his master's voice. There was no doubt in his mind that Mack was down there.

'I'm goin' for 'elp,' he shouted, leaving his stick in the ground as a marker, and hurrying back to the pub.

The mountain rescue team was summoned, being familiar with the terrain and having all the right equipment to implement an underground rescue. Whether it be people, dogs or even sheep, there seems to be no scenario that they haven't encountered. So theirs was the obvious number for Alec to call. We have, over the years, had a good many cragfast sheep, stuck on ledges and precipices, unable to extricate themselves from their predicament.

'I've a dog down a mine shaft,' said Alec. 'Does ta' think you lads can ger 'im out?'

It wasn't long before a mountain rescue Land Rover pulled up outside Tan Hill.

'Follow me, I'll show yer where he's at,' said Alec, who'd been pacing back and forth waiting for them. Back into his van and down Arkengarthdale he went, with the Land Rover behind, until he pulled off at the side of the road.

'We'll 'ave to walk frae 'ere,' Alec said.

The mountain rescue team unloaded their pot-holing equipment, ropes, helmets and lamps and pulled on their suits, then set off to the hole. The first job was a preliminary examination of the scene, to decide how deep the hole was and whether there was any danger of the ground surrounding it collapsing.

'What 'ave we got 'ere, then?' said one of the team.

'It's an owd ventilation shaft outta t'coal mine,' said the team-leader Pete. 'Look see, it's all walled around the sides.'

'Should've 'ad a bloody lid on,' said Alec.

'It's a wonder that thee didn't ga' down it an' all,' said Pete as he shone the flashlight down into the bottom. 'It could be an 'undred foot to t'bottom.'

'Can yer see Mack?' said Alec.

'No. I can see watter, but nae dog,' said Pete. 'Let's be goin' down an' 'avin a look.'

A metal tripod arrangement that supported a pulley was assembled and placed over the shaft; then the smallest member of the team, a woman, was harnessed up and slowly lowered down.

Minutes later, she reported back that the dog was there, but he was in a bad way. A decision was made: Mack needed to be brought to the surface as quickly as possible. She took the dog in her arms and the team pulled them both back to ground level. The cold, wet broken body of the bedraggled dog was laid on the heather and the team gathered round. As Alec knelt beside him, Mack slowly turned his head to him and closed his eyes as the life ebbed out of him.

'We tried,' said Pete.

'You've done good,' said Alec, sighing and looking at Mack. 'There's nowt more we coulda done to save 'im.'

It was a wonder that Mack had survived the fall at all, but at least Alec now knew what had happened. It's far better to know the outcome than to spend sleepless nights always wondering.

The very next day the coal board sent a man out to inspect the scene, and within days the shaft was filled with stone and made safe.

Mack was brought back to Tan Hill, and buried near the Pennine Way.

Plenty is known about the coal mine at Tan Hill, but I had no idea until recently that there were also private mines, sunk by farmers to take coal out of the ground for their own use. A visitor, who was on holiday in the area, told me he had once lived at Hill Top, the house at our road end. Now just a house, it had once been a small farm with a couple of barns and a few fields which we now farmed. He reminisced about the bitter winter weather he remembered, and how his new home on the south coast is much warmer.

'Nay wonder they 'ad to 'ave their own coal mine,' he said.

'Eh?' I said. 'Who had their own coal mine?'

'Hill Top,' he said. 'Damned shame when they privatized the coal mines, it 'ad to be blocked up.'

It transpired that the rough old patch of moor behind the sheep pens was the grassed-over remains of the earth workings surrounding a coal mine. It was only a small seam, and the coal was of poor quality – but it was enough to supply the farm and warrant construction of a primitive underground railway track, along which wagons loaded with coal were pulled to the surface.

Reuben had been up at Hill Top only weeks before and told me how he'd found some long pieces of iron alongside what looked like foxholes. I had to conclude that he'd probably been very near to the entrance to the mine. I decided not to enlighten Reuben, in case he decided to try coal mining for himself – he's an inquisitive and adventurous sort. He recently discovered a cave near there, but when I asked about it he told me it was inaccessible to someone of my dimensions.

'What does that mean?' I asked.

'Mam,' he said flatly, 'yer couldn't get yer arse in there.'

He doesn't mince his words. But when he took me to see the cave entrance, I could see he was right.

Working dogs, unlike pets, are exposed to dangers like screes, rocks and rivers, so sadly they are occasionally lost in the line of duty. It's a hard life for both dog and shepherd, and there are many stories of faithful sheepdogs committing acts of bravery and supreme devotion by staying at the side of an injured shepherd, providing comfort and warmth or even summoning help. There are far fewer tales of sheepdogs causing accidents and then blithely abandoning their shepherds – but we have one.

One lovely autumn day we were going about our business, amongst the sheep. The children and I had driven the Land Rover to the Hill Top fields to ruddle a tup, taking Kate to gather up the yows. We were in no hurry, as the children were wrapped up warm and were happily poking around the barn, climbing the outside drystone-walled staircase, jumping from it into the small grassed-over midden. The majority of the yows had been tupped by now, so Clive was at home in the sheep pens smitting a few of them. He had finished his job and was returning them to the field when they decided to take a detour. Instead of turning sharply to the left after he opened the pen gate and streaming across the little packhorse bridge and into the Big Breas, they galloped straight down the yard in front of the house, heading for the cattle grid and the moor.

A cattle grid is designed to prevent livestock from getting across, but for some of our wily old dears it is no hindrance. They accelerate to full speed, take an almighty leap and clear it with room to spare. The problem comes when the other sheep following them are not so skilled at the long jump, but still attempt it just because they can see their friends at the other side. It's their lemming tendency.

This detour the sheep were taking was potentially disastrous, as misjudging the leap could result in broken legs. Clive set off after them on the quad bike, cussing and lamenting that I'd taken Kate, the faster dog, with me. He came to an abrupt stop outside Bill's kennel, jumping from the bike to unchain Bill – who was already in a frenzy, having heard the revving motorbike and the accompanying shouts. Released from his shackles, he put in a typical excited dog move, spinning around while waiting for instruction from Clive. Clive's mind was now on heading off the sheep before they reached the cattle grid and he'd forgotten to put the brake on the quad bike, which was now slowly rolling down the yard.

Bill somehow got himself between Clive's oversized thermal moon wellies, tripping Clive, who fell headlong, arms outstretched. The bike was still rolling, and Clive cracked his forehead on the metal rack on its rear. He hit the ground hard. The bike carried on, eventually stopping when a picnic bench blocked its path. Clive, meanwhile, was laid flat out on the concrete. Bill sniffed at him, and then quickly lost interest. By the time a group of passing walkers found a dizzy, disoriented Clive sitting in the yard, Bill had left the scene of the accident and was quietly stalking hens. So much for devotion and loyalty – he was making the most of his free time.

It's probably quite lucky that Bill was distracted, because Bill cocks his leg on anything and everything, and Clive's inanimate form might have been tempting. I was oblivious to all of this, in the field at Hill Top with the children. The walkers helped Clive to his feet and led him into the house. He had a nasty, deep cut on his forehead, with blood trickling down his cheek, and another cut under his eye. When I got back I was met by two of the walkers at the door.

'There's been an accident,' they said. I pushed past and found Clive holding a towel to his head.

'We didn't see what 'appened,' said one of the ramblers. 'Just found 'im in t'yard.'

I thanked them, and then quizzed the pale-faced Clive.

'I fell over mi dog,' he said. 'It's nowt, I'll be fine.'

I looked at the damage and then broke the news that he needed to be patched up properly. There wasn't any sticking plaster that was going to cover wounds this big.

'I've got so much to do,' he said. 'I ain't got time for no blinkin' 'ospital. I need an 'ospital trip like I need an 'ole in the 'ead.'

'Yep, you've certainly got one o' them,' I replied. Luckily the doctor at our local surgery volunteered to do the stitching, and with the help of a no-nonsense nurse and a good length of suture Clive was soon patched up. It had shocked him and he was quiet for the rest of the day, waking up the next morning with a shiner of a black eye.

The doctor did a marvellous job, and within weeks it had all healed up very nicely. Clive looked in the mirror and frowned.

'If mi scar had gon t'other way I'd 'ave looked 'ard, like a gangster,' he said. 'But as it is I've just added another furrow to mi brow.'

12

December

December is officially the first month of winter, but up here in the hills we will have been feeling the cold for some time. In some years we have had a light covering of snow on the ground when we've been tonsing the tups ready for sale in late October, and small packets of compacted snow can survive under rocky overhangs at the moor until late May.

The sheep are back on their respective heafs at the moor by now, and we supplement their diet with hay and sugar beet pellets.

By December my aim is to have the dairy well stocked with provisions for us, and the meal house packed to the rafters with enough food for the animals to see out a storm.

Usually we have a barn full of year-over hay – hay that has been left over from the previous year. Time doesn't do it any favours, but in a bout of bad snowy weather it can be fed to the animals when the pangs of hunger bite. Sheep usually prefer not to eat the clumps of seaves, but when these are the only greenery, their stems poking out from the snow, they will nibble them. Throwing out canches of the 'storm hay' or, in a bad time, straw, will fill their bellies and they'll ride out the storm, especially in the early winter months when they are only in the earliest stages of pregnancy.

We try to preserve the smaller-sized conventional bales for feeding the flocks at the moor. For any sheep in the barns and the horses we use larger round bales of haylage, some of which we've made on the awkward steeper areas of the hay meadows, and some that we've bought in from any local farmers who have a surplus. One year, Clive bought a trailerload of round bales for 'handy money' and set me off on a tractor to Kirkby Stephen to bring them home.

'If yer get yerself over to Kirkby this mornin' then there'll be someone about to load the bales onto yer trailer,' he said. 'I'm too busy to do it, an' yer can pick up a bit o' shoppin' whilst yer there.'

True, I did need a few bits and pieces from the supermarket, but it is no mean feat manoeuvring a tractor and a twenty-five-foot trailer into the bays of the car park. The previous winter we'd bought loose clamp silage for the cows and I still remember how the top layers of the wedges of grass, pickled and pungent-smelling, blew off the trailer and across the car park. Shoppers laden with bags and pushing trollies were wrinkling their noses and brushing from their hair and eyes the wisps of chopped grass that had been carried on the strong breeze, while I, clearly identified as the culprit by my wellies and waterproofs, turned a blind eye to the complaints and carried on shopping.

Dressed warmly, I walked down to the other side of the bridge where the tractor was parked. The trailer had been abandoned on the hard standing beside the river. Hitching up a trailer to the tractor is not as difficult as hitching up a trailer to the pickup or Land Rover, as you can see clearly from the back window of the tractor whether you are lined up right to drop down the tow ball and link up to the trailer drawbar. I did this without a problem, then climbed down and connected the hydraulic brake pipe and the light electrics. Unfortunately I'd overlooked the fact that the hydraulic pipe was dangling loosely, and as I turned the tractor quite sharply towards the road and pulled away, the pipe caught in one of the tractor's wheels, stretching it until it snapped. I didn't notice any of this until I turned round in my seat and saw the pipe trailing on the ground.

Oh, hell, I thought. *Clive is not gonna be reet impressed when 'e sees what I've done.*

In that split second I decided to put off saying anything to Clive, as we couldn't repair the damage ourselves, snapped hydraulic pipes being something that only Metal Mickey could sort. Time was of the essence here. Reaching into my pocket for a strand of baler twine, I tied the snapped pipe back to the trailer and tried to remove the other half of the pipe from the connector on the back of the tractor. I'm not a weakling, but the coating of oil on my hands, and the fact that the metal connector had been bent by the pressure of being pulled at by the force of the wheel, meant that it would not move. Cussing to myself, I reasoned it would just have to stay attached – after all, it wasn't going to interfere with my job. The big problem was that the trailer brakes were obviously not working, but I figured that, as long as I drove slowly and carefully, I'd be all right.

Off I went. It had begun to rain, and the clunky, unrhythmic wipers were smearing the blobs of muck that had crusted on the windscreen after the previous week's muck-spreading. I drove as quickly as I could, meeting no vehicles coming the other way and never taking my foot off the accelerator. Unladen, the trailer was no bother when it came to stopping, but when the bales were on board it would be a different story. I daydreamed a little, mulled over what I needed to get from the supermarket and looked across to the open moors, studying our neighbours' sheep.

Reaching the top of Tailbrigg, I was brought back to reality as I rounded the bend and saw a car approaching up the hill. Needing to give him some room so that we could pass each other, I put my foot on the brake and began to steer towards the crash barrier. The tractor didn't respond the way I expected: in fact, it didn't really respond at all, the steering feeling heavy and the brakes spongy. I tried to change gear but couldn't depress the clutch pedal. I frowned, grasping at the steering wheel and pumping the brakes more forcefully. By now the dashboard was lit up like a Christmas tree, flashing warning lights, and an alarm was ringing from somewhere near the ignition. I began to panic. Taking my foot off the accelerator had slowed me down fractionally, but I was still in top gear and travelling far too quickly for a descent down a 1:5 gradient hill. Even more worrying, the chap in the car coming towards me seemed to be counting on me swerving to one side at the last moment. The steering wheel just wouldn't budge. I hit the horn, hoping that the oncoming car would realize that I was on a runaway tractor and trailer. Turning round in my seat to see what the trailer was doing, I saw a jet of oil spraying like a geyser straight up into the air from the back of the tractor.

There was only one thing for it. Pulling on the handbrake I made an abrupt emergency stop like no other. I quickly switched the engine off. I was grateful and very lucky that I had one of my legs braced against the foot pedal and was half twisted backwards in my seat, for this is what saved me from hitting the windscreen face first. Fortunately the car coming towards me had also stopped. I sat still, exhaled loudly and willed my hands to stop shaking. Taking a moment to compose myself, I looked at the road and saw rainbow colours in the film of oil that was slowly spreading across the damp tarmac.

I climbed down from the tractor and went to investigate what had gone wrong, and to apologize to the other motorist. There was an oil slick from the top of the hill, where I'd first tried to brake, right down to where I'd come to an abrupt halt – there was gallons of it. I assumed that there was no oil left in the tractor. The other motorist was very understanding, until I restarted the tractor and tried unsuccessfully to move it out of the way. It moved neither forwards or backwards. Stalemate. A line of cars had appeared behind me, and the drivers were none too impressed with the situation.

'I could do wi' borrowing someone's mobile phone,' I said.

I rang Clive, but of course nobody answered. I left a message and hoped that he would respond.

Respond he did, and it wasn't pretty.

'You've done what?' he shouted loudly, as I held the borrowed phone away from my ear.

He appeared half an hour later in the pickup, children in the back, bringing a random selection of spanners. By this time I'd been deserted, everyone else having turned their cars around and either abandoned their plans for the day, or found an alternative route.

Not much was said. Clive shook his head as he wrenched the hydraulic pipe connector out of the fitting.

'There,' he said, as he ceremoniously handed it to me.

'Sorry,' I muttered. It seemed obvious in hindsight that leaving the snapped pipe attached would mean that the valve would be open, and free to spew oil from the tractor until no more remained.

Between us we carefully tipped three-quarters of the contents of a very large drum of oil into the back axle, and it wasn't long before I was mobile again. We went our separate ways. The rest of my journey was relatively uneventful, and I was soon back at Ravenseat with the bales.

Shortly after this, my thoughts turned to Christmas and shopping for the festive season. I don't believe in pointless presents, preferring to give each of the children a carefully chosen and appropriate gift. But my idea of 'appropriate' doesn't always tally with everyone else's. Geoff, our knacker man, turned up just before Christmas a couple of years ago to collect the body of Van Gogh, one of our veteran tups, who had succumbed to pneumonia. I went down to the bridge to pick up the paperwork, and to give Geoff a box of chocolates and a Christmas card.

'How yer doin', Geoff?' I said. 'And 'ow's Donna? Looking forrard to Christmas?'

'Aye, I's champion. But I's not wi' Donna any more, I's back wi' Katy.'

I decided not to delve any further into his love life.

'Anyway, there's sum chocolates 'ere for yer.'

'Ta varry much,' he said, putting them on the front seat of the lorry. 'Ah've got summat for thi kids, an' all.'

'Oh, you shouldn't 'ave,' I said, smiling.

'I'll just get it, it's in t'back of t'lorry.'

Alarm bells started ringing. What sort of Christmas present was delivered in the business end of a knacker van? I soon found out.

'Aye, I's gotten a dead reindeer for 'em. Just picked it up this mornin'.'

A dead reindeer! Crikey, here's me trying to get the little ones to recite 'Prancer, Dancer, Donner, Blitzen . . .' and here was one of them (Vixen, maybe) laid out cold in the back of Geoff's knacker wagon.

'I's thinkin' they'd mebbe like to see it an' saw off t'antlers . . .'

'Errr . . . nooooo, Geoff, I don't think so,' I said, contemplating how that sort of traumatic memory would be associated with every Christmas-time thereafter.

Admittedly Reuben does have a collection of horns and antlers in the woodshed – which isn't really that macabre, because there's money to be made from selling them to stickmakers who will bend, carve and craft them into decorative handles for walking sticks. But hacking the antlers off a reindeer at Christmas just seemed wrong. This was one set I decided they could do without.

I collect most of the children's presents throughout the course of the year, so my Christmas shopping is just about getting wrapping paper, cards and crackers: nothing too taxing. Not long after the runaway tractor incident I was asked to say a few words at a tourism conference in Harrogate, and decided this would be an ideal time to get these bits and pieces.

The day started badly, as I was already het up by having a cow to calve that morning. The cow and calf were fine; the calf was just a bit bigger than normal, but it meant I was running seriously late. I wasn't sure how long it would take to get to Harrogate but decided that I would put a smart dress on and sort out the finishing touches – hair, make-up, tights and shoes – when I got there. I took Annas with me, while Clive kept an eye on the bigger children.

'Don't panic,' said Clive. 'Stay calm.'

This was guaranteed to make me panic, and did nothing to calm me.

After slinging a variety of bags into the pickup and strapping baby Annas into the car seat, we were off. The first thing I noticed was that the footwell seemed wet. No matter, I thought, working vehicles are usually damp. No farmer ever uses a boot bag or changes into driving shoes, so the footwell was normally disgustingly wet and muddy. There were a few dog hairs, too; not surprising really as Bill the sheepdog doesn't like travelling and only tolerates a car journey if he can curl up on the passenger seat so no wonder there was a pervading smell of wet dog. A quick spray of perfume would soon rid me of the whiff, I thought. Oh, and one of the leaf springs under the pickup's chassis had recently broken so the whole motor rather tilted to one, the back door had become detached and was officially classed as missing in action, and a pair of faded red fluffy dice that the children had found at Tan Hill Show hung from the rear-view mirror.

I hadn't gone far down the dale when my foot started slipping off the accelerator. Glancing down, I realized that the small pool of water that had accumulated in the footwell was actually oil, from the oil-drum behind the seat which Clive had brought to rescue the tractor. It had tipped over, spilling the remains of its contents. When I went uphill the oil disappeared under the driver's seat and into the back, returning to the footwell when I went downhill.

'Thank goodness I didn't put on mi heels,' I thought, as it was bad enough getting a foothold in my wellies. A couple of hours later I was negotiating Harrogate's one-way system, with one eye looking at a map printout that was laid on the seat. I kept glancing towards the clock. Annas slept in her car seat, without a care in the world, her little rosebud mouth pursed, sighing occasionally and turning her head without opening her eyes. After a good while stuck in traffic and looking for a pub called The Stray, where I was supposed to take a left, I stopped to ask for directions. Apparently Harrogate had no establishment by that name, The Stray being a stretch of open parkland.

I finally pulled into the gravelled car park of the country house hotel with twenty minutes to spare. Only one parking space remained, my dilapidated pickup looking somewhat incongruous between a yellow Porsche and a brand new Range Rover. The great and good of Yorkshire were all making their way to the grand foyer, while I rummaged around gathering up bags of clothes and shoes. I had left Ravenseat on a typically frosty cold morning wearing my camouflage army surplus coat over my dress, and now, in temperatures a few degrees higher, I looked seriously overdressed. Annas started to squawk. I picked her up along with the assorted bags, and did what could have been mistaken for a war dance on a neatly manicured lawn in an attempt to wipe off the oil from my wellies.

I needn't have worried about the impression I made in my strange outfit, because the hosts of the event were typically welcoming and uncritical.

'Ah, Amanda . . . glad you could make it,' said one of the organizers, vigorously shaking my hand. I shrank away, worried what state my hands were in. Rough and shovel-like is normal for farmers. I began to babble:

'I'm sorry I'm late . . .'

He cut in: 'Yes, yer need to change, I can see that. Follow me to t'green room. We've laid out some breakfast for yer, too.'

Things were looking up. I followed him up a winding back staircase away from the large crowd of people who had assembled, chatting to one another or networking, depending on how you see it. I was ushered into a small dressing room and he switched on the lights. Annas's eyes lit up nearly as brightly as the bulbs around the Hollywood-style mirror. I squinted at my reflection, rubbed my eyes and ruffled my hair. I looked ropey, to say the least.

'I'll leave yer to it,' he said. 'There's some breakfast in t'green room, just along t'corridor. Someone'll come an' get yer when it's time for you to go to your seat.'

I thanked him, and proceeded to tidy myself up as best I could. Then, tempted by the prospect of breakfast, I went along to the green room. Annas was getting hungry, chewing her little hand, and I was hoping there might be a yoghurt for her.

I was not disappointed: juice, fresh coffee, fruit, pastries and yoghurt. My attention was momentarily diverted from the food by a man in a suit and maroon tie sitting on a small upright sofa in the corner. I smiled and said hello, then went back to filling my plate. I plonked myself on the other end of the sofa and put my overloaded plate on a small table. Sitting Annas on my knee and tucking a serviette under her chin, I began spooning yoghurt in. She beamed, yoghurt dribbling down the sides of her mouth.

'What a morning I've had,' I said, attempting to engage with the man at the other end of the sofa.

He nodded, saying nothing. I guess it must have been my nerves kicking in because I started to rabbit on about cows calving, not having time to get changed, the children, trouble with the pickup, you name it, ending with how I found myself sitting in the green room at a tourism conference. Throughout the whole episode he uttered not a word, no outward sign of emotion. He didn't look interested, or even disinterested. He passed me my cup of coffee from the low table.

'Anyway,' I said, in full spate. 'If this whole conference thing gets too boring an' Annas cries, then I'll just plug her in, give 'er a good feed an' then she'll be fine. These things can be sooooo dull, dull, dull.'

I somewhat emphasized the 'dull'. Annas finished the yoghurt, I polished off the pastries and then it was time to go. I was placed in the front row of the auditorium and was told that the presenter would call me onto the stage. I was slightly flustered, as I'd decided to just ad-lib, rather than write any notes. Annas was sitting on my knee and she too was fidgety, picking up on my nerves. So I plugged her in, covering her little mop of sprouting blonde hair with an oversized scarf that I'd had the sense to bring with me. Breastfeeding has only once caused me embarrassment. Usually when I'm on the farm or out and about in the fields it simply means there is a ready supply of warm milk for the baby. The folks at the auction treat it with casual indifference. It amuses me when I hear of women in towns and cities being criticized for feeding their babies in public: at the auction mart, a microcosm of traditionalism and convention, it is accepted without a second glance.

'Nowt like a sooked 'un,' farmers say.

Only on the one occasion had I been left red-faced. I was walking down a street, minding my own business, feeding the baby as I went along. If I have a baby in the front papoose it is possible to shuffle her into a feeding position, leaving my hands free to lug shopping bags or push a trolley or wheelbarrow. A lady, who I assume was a tourist, stopped me and began to talk: 'Ooooh, it *is* you,' she said. 'The lady off the TV, an' you've got another baby.'

Before I had a chance to say anything she reached forward, pulling the edge of the papoose back, and peered in.

'I love babies,' she went on, then stroked what she thought was the baby's head. 'No hair?' she said. 'A baldy.'

'Not quite,' I said, pulling away and shoving my booby back into my shirt.

The conference was in full swing, Annas was quiet and I was on tenterhooks awaiting my curtain call. Suddenly the lights dimmed and a big screen lit up on the stage behind the lady presenting the event. Dramatic music thumped through the loudspeakers, then my own face appeared up on the screen. From out of nowhere had sprung a cameraman, camera on shoulder pointing right at me and wielding a fluffy microphone that he thrust in my direction. I half-smiled, half-grimaced, then all went quiet.

'Here we have Amanda Owen,' said the presenter. 'Can you come up onto the stage, Amanda?'

The camera followed me, projecting my image, complete with audio, onto the big screen for the audience's enjoyment. To my embarrassment, very loud suckling noises were broadcast across the packed room. Sitting myself on the couch on the stage ready for a brief interview, I managed to get Annas unplugged. Her little goldfish-like mouth was still sucking, but now just fresh air, as she emerged blinking from under my scarf.

I did my little talk, calming down a bit. Annas even began to enjoy the limelight, giving big animated smiles. Then it was time for the next guest to come on stage. I sat still while the 'mystery guest' got a great build-up.

'We have a special VIP guest, who we are so lucky to have with us today,' said the presenter. 'He's known to millions . . . a familiar face to us all.' She babbled on, enthusiastically.

Onto the stage strode the nondescript man in the maroon tie who I'd met in the green room. Riotous applause filled the room. I frowned – *known by millions*, but not to me . . . I stood up, beamed and shook hands with him while struggling to identify him.

'Ladies and gentlemen, I give you the deputy prime minister, Nick Clegg,' said the host.

I was mortified. I'd been trapped in a room with the deputy prime minister, and I didn't even know it. I've never taken much interest in politics, although there are certain faces I'd know: Eric Pickles – no idea about his policies but I'd know his mush if I saw it; Boris Johnson – I'd know his hair anywhere; and our local MP, William Hague – I'd know his lack of hair anywhere.

I once stood on a balcony alongside him and lamented the proposed loss of our nearest maternity unit. It still closed, but at least we tried to save it. I remember looking at the sea of people in front of the county hall and commenting that there were a few placards bearing anti-Hague slogans among the crowd. Stupidly I then mentioned a grassy knoll, and before I knew it burly security men had jostled into position between William and the crowd, presumably ready to take the bullet. I decided from then on to keep my mouth shut and steer away from all things political.

I didn't hang about after the conference. I was off my home turf, which always unsettles me, so after picking up the bare necessities in Harrogate I was soon back en route to Ravenseat. Not long after, the pickup went to the garage for some running repairs: the oil remained in the footwell.

'Weell, look on't bright side – it'll nivver rust, Mand,' said Clive.

On my next foray into civilization, after little Clemmie was born, I decided to take Clive with me, in the Land Rover this time. Clive hates the Land Rover almost as much as he hates shopping, but on this occasion it was vital he came along, as he needed some new jeans before the Christmas parish party. Typically, once again the cow that we'd been watching intently as she warmed to calve had decided today was the day, so after doing the bullocking up around the yard we oversaw the birth and then set off with Sidney, Annas and Clemmie on the back seat. Clive complained for the entire hour-long journey to Kendal.

'No cup holders in here, Mand,' he said.

'Yep, I know that,' I said. 'But it's not like we've gotta cup to put in one.'

That was just the start of the complaints.

'Air conditioning,' he said, pulling the knob on the front vent that opened an exterior flap. 'Bit basic, in't it?'

'I like simple and basic,' I retorted. 'After all, I like thee.'

Once we were on the busier main road he muttered, 'I can't 'ear meself think. Can't you go into a higher gear?'

'Yer what?' I said.

We were later than we'd planned, and Kendal was busy. I hadn't been for years and, flummoxed by the traffic, I looked in vain for a parking space.

'Multistorey,' said Clive, pointing to the left. 'Yer need to change lanes.'

A motorist flashed his lights and let me in; I indicated left, and edged forward towards a slip road and the multistorey. It was another tight turn to the left, and then into the building – or it would have been, if the Land Rover had fitted. As we went under the overhead barrier I actually ducked, even though I was in the driver's seat. There was a loud bang, and when Clive got out to assess the situation, the long yellow-and-black-striped bar was still swinging back and forth from the impact.

'Just keep gaan',' he said.

'I'm not gaan' any farther, I'm gonna be scrapin' t'roof if I do,' I said. 'It'll be like a blinkin' dodgems, there'll be sparks flyin' all roads if I go in.'

'Well, do summat,' he said, aware that there was a sizeable queue of cars now lined up behind. 'Mebbe if I let some wind outta t'tyres . . .'

'Aye, an' then what?' I said. We were in danger of having a full-scale domestic right there in the middle of the road.

I backed up, and so did everybody else, as they had to let me out. Finally we decided that we'd park at a supermarket and walk into town. It didn't take long to get Clive kitted up: he was very focused, knew exactly what he wanted and didn't dally. I told him that I'd go and get us all a drink from the coffee shop on the high street.

'I'll stay 'ere an' watch the world ga by,' he said. 'An' bring us a sausage roll, an' all.'

He settled himself on a bench in a pedestrianized area with the children. Sidney and Annas just messed about, getting in the way of shoppers and pressing their faces up to the shop windows, everything new and intriguing to them. Clemmie was fast asleep in the papoose, so I handed her over to Clive without her waking up and she lay quietly, nestled securely in his arms. Clive doesn't mind looking after the children, positively revelling in the attention it brings him. Just leave him with a couple of the children and a baby, and women flock to him in droves.

'It's just like a stampede of coos comin' at me,' he says by way of feeble apology when I return to find him surrounded by a squadron of ladies.

'Once that babby bawls out like a calf then they all come, I couldn't hope to fend 'em off.'

As it happens, he never did get the coffee, as I got waylaid: distracted by a junk shop, set back from the main street, with a display of glazed chimney pots and stone troughs outside on the pavement. Peering through the mullioned windows, I could see an array of miscellaneous objects ranging from slightly unusual to the downright weird – military hats; a very large case of stuffed sea birds; oil paintings; and what I thought was spiked leather fetish gear, although it turned out to be pairs of old-style mountaineering crampons, all strapped together. (Clive said this showed how my mind works.) Among the ephemera were two large iron wheels which had perhaps once been part of a horse-drawn hay rake, or maybe even a shepherd's hut. I had to buy them: Reuben would love them, they'd likely come in useful one day, and in the meantime they'd look nice on display in the farmyard. We hag-gled a bit, and eventually a deal was struck. Then I discovered just how heavy the wheels were, and remembered how far away the Land Rover was. At least, as they were wheels, I could roll them – and by a combination of lifting, pushing and guiding them in the right direction, I was soon back at the precinct. I swear I saw Clive groan as he spotted me coming.

'A roll,' he said. 'I wanted a sausage roll, not a pair o' wheels to roll.'

Sidney liked the wheels, and even Clive admitted to liking them – but not until very much later that day. We were both sweating by the time we got them back to the supermarket car park. I picked up the Christmas essentials with the children dancing their way up the aisles, full of excitement at the impend-ing visit from Father Christmas, while Clive recovered from the ordeal in the cafe.

One year I asked the children what they wanted for Christmas, and the answer was 'Pet rabbits.' I was amazed. There are hun-dreds of rabbits in the fields, easily spotted lolloping around at dusk, eating our precious grass. Forget *Watership Down* and Bright Eyes: Clive actively encourages the terriers, Chalky and Pippen, to try to catch themselves a furry feast. We've occasion-ally caught them ourselves, and I've made them into pies.

I need more animals to care for like I need a hole in the head, but I knew the children would be super pleased if we got them, so I wrestled momentarily with the dilemma: 'Shall I be cur-mudgeonly, horrid mum, and say no, or should I just go with it?' I'd spent my childhood asking for all manner of pets, and being turned down flat almost every time. As a result, I was guilty of letting my heart rule my head when it came to taking on unsuitable or impractical animals once I'd moved away from home and had my own little cottage at Crosby Ravensworth. I'd had pet lambs that cost me a fortune to rear on powdered lamb milk (when I hardly had enough money to feed myself) and a billy goat who smelt terrible, and who frequently broke free from his tether and ate people's washing.

So it came to be that our barn loft became home to seven very sweet Dutch rabbits. The children brought them down into the house to play with them, which perplexed Chalky and Pippen, who had to be corralled safely away. One minute Clive was telling the dogs to see off the rabbits, the next the children were stroking and cuddling them . . . what's a terrier to make of it?

It must have been during one of these playtimes that the Ravenseat rabbit-breeding programme was inaugurated. One morning Reuben rushed into the kitchen and announced: 'You know them two boy rabbits that Miley and meself 'as?'

I guessed where this was going. 'Barry and Gary?' I said.

'Aye. Well, there's two baby rabbits in t'hutch with 'em this morning.' He was bursting with excitement.

Clive, who was leaning against the Rayburn warming his hand round a mug of tea, smiled wryly as he looked at my swelling belly.

'Aye, they say they breed like 'umans . . .'

In December even the barns look festive, with holly hanging from the rafters and beams above the cow byres – although this is the old-fashioned way to ward off ringworm, rather than welcome Santa.

The annual parish Christmas party is a chance for the children to let their hair down, take part in riotous games with other children and meet the great man himself, Santa. Getting eight excited children into their party clothes and to the church hall is a feat in itself and I'm ashamed to say that this year we were a little late and missed the start of the party. This had nearly disastrous consequences when, as I pulled up outside the hall, the beam of my headlights caught a startled half-dressed Santa. He was hopping about on one leg, his red trousers around his ankles as he struggled to pull them up over his jeans and wellies. He was last seen trying to regain some composure, realigning the beard and hair ensemble before he lurched off into the darkness behind the church wall. I quickly switched my headlights off, but I feared it was too late, the damage already done. Santa's cover was blown.

'Wasn't that Matthew frae't Hoggarths?' said a perplexed Reuben.

'Shut up,' I hissed at him while Raven laughed uncontrollably.

Nothing more was mentioned about Santa or his alter ego and the children spent the evening stuffing themselves with party food that folks from up and down the dale produced. I go to great lengths to make sure that our edible offerings are up to the mark. There's nothing more depressing than loading your uneaten cakes and buns back into the tin afterwards, or hearing my children trying to persuade their friends to eat our food. There have been times when I've overheard the children's conversations and it's all sounded very cheffy.

'Trifle was nowt,' one of my brood will say, as they bounce about in the Land Rover on the way back home.

'They can't slother thi' trifle in booze like we do,' I'll say. Maybe I should be worried that they've got a taste for the finer things in life . . .

On Christmas Eve it's off to our friend Elenor's house for the Christmas carol service. Then home, the remainder of the day spent with the children frantically rewriting Christmas wish lists and sending them up the chimney in some last-ditch attempts to communicate with Santa. Then before bed they try bribery, with the ceremonial setting-out of mince pies, booze, carrots and a hay net for Santa and the reindeer. After every bedstead has a stocking tied to it, it's time for sleep – but it can be a long time before the excited chatter from upstairs dies down.

Every year, the final instruction is the same: 'Wake us at midnight so we can go to t'stables.'

Sometimes I do, other times I alter the kitchen clock so that we can get to bed at a reasonable time. I always like walking across to the stables on a clear, crisp night, the yard illuminated by the moon and the night sky a mass of twinkling stars. The little ones are invariably asleep by the time we call them, but the big ones, wrapped up well with kitles over their pyjamas, hats down over their ears, always hope to find the horses kneeling in honour of He who was of a stable born.

We try to get to the stables without alerting the horses, to catch them unawares so that they don't feel the night's cold after the warmth of their beds. Only once have we managed to get to the stable door without their super-sensitive ears hearing us approach. Once the first nicker is sounded, there's a chorus of neighs to welcome us. The story is the same, every year. The young horses turn their heads and shoot us supercilious looks, questioning why we are invading their warm and peaceful stable. Little Joe, the veteran Shetland, who is now partially deaf, will be standing with his legs locked, his head bowed whilst dozing. Dreaming of what? I'm not sure, but I believe horses, like dogs, dream. I used to watch Deefa, my very first sheepdog, curled up amongst the cushions beside the fire, dozing when she felt relaxed and safe. After a while she'd emit little yelps and her tail would feebly half-wag while her legs pistoned back and forth, dreaming of chasing sheep or rabbits. The horses, who also have memories, may dream of the heady scents of summer meadow grass, or perhaps the tantalizing sweetness of a freshly picked apple.

I am overwhelmed with feelings of loss and grief when I see one of the stables, Meg's, standing empty. Her blanket still hangs from the beam, gathering dust; her feed bucket still sits in the corner, an indentation in the side where she'd impatiently kick it. Her bridle, saddle, breastplate, numnahs, brushes, fringes, every mortal thing that a horse could ever want, though likely never needed, still sit in the porch. All of our horses have their foibles and idiosyncrasies, and Meg was no exception. She pulled contorted faces every morning, furling her lips right back, particularly if she caught sight of her arch-enemy and nemesis, Queenie. Meg was a character, strong-backed, sturdy and sweet-natured (Queenie would beg to differ on this), and we conquered everything together. We'd shown and won championships, and it was always a two-way thing: two hard-headed women (literally hard-headed in my case, after she bolted and I fell off her). She was as happy to shepherd the sheep as she was in the show ring. I trusted her implicitly and she never wavered, never let me down.

This was what I thought about when, after spending almost twenty years together, I looked for the glint and flash in her eye and saw instead dullness and pain. She'd survived when the odds were stacked against her, enjoying one last summer at Ravenseat, wandering wherever her heart desired – usually through the best hay meadows, much to Clive's annoyance.

Seeing this finest of horses fading away in front of my eyes was very painful, and the time eventually came to call an end to her silent suffering. I tried to talk myself out of it, but knew that I'd explored every avenue. Old age, illness and infirmity had caught up with my beloved Meg, and she needed me to give her the dignified end that she so deserved. We buried her on the hill

overlooking the farm, her name etched in stone, forever preserved.

'Meg, a finer nivver lifted leg.' That's her epitaph.

There's no time to be maudlin. The cycle of life continues,
and we now have Meg's daughter Josie and granddaughter Della
with us at Ravenseat.

Jobs need to be done on Christmas Day just the same as on any
other day. We don't mind: for us and farmers everywhere, the
work goes on, and the children are used to this. It doesn't mean
that we don't have fun, and Santa, of course, visits all the animals in the farmyard: the cows in the byre, the sheep in the shed
and the dogs in the kennel. The dogs get a Christmas dinner
from the turkey carcass, but maybe not for a few days, as the
supersize mutant turkey takes a bit of wading through.

The cows get salt licks, and they happily spend many hours
curling their long, abrasive tongues round the jagged pinky-
white lumps until they are as smooth as pebbles. The sheep also
enjoy mineral feed blocks, which come in two sizes: small rect-
angular ones for smaller flocks, or giant, heavy round ones for
where there's a lot of sheep and a lot of competition. We're keen
to give sheep an extra boost through the hardest winter months,
and feed blocks seem to be the answer. We put them out on the
heafs and they encourage the yows to stay in the vicinity, which
helps us when we check on where they are.

For quite a few years the poor old sheep were given the same
mineral feed blocks for Christmas – not the same type, but the
same actual blocks. For some reason they didn't seem able to
make a dint in the glistening black surfaces. It had cost quite a
lot of money to buy blocks for all the different heafs of sheep,
but they didn't seem to be getting much benefit from them, for
all their persistent licking and even gnawing (and this worried

us, for a sheep is only as good as its teeth). I attacked one of the blocks with my penknife, to try and break the surface and let the yows in, but they were so rock-hard that I was hardly able to dig my knife in. Reuben decided on a more drastic solution: a blow torch. The surface liquefied and became molten molasses, but as soon as it cooled off it solidified again.

When the feed rep came into the yard, he got a warming from Clive.

'Christ, were these feed blocks med at Willy bloody Wonka's factory?' he said.

'Sorry, I's not followin' yer,' said the poor rep. 'What seems to be the problem?'

'They're bloody everlasting,' said Clive. 'They last forever an' a day.'

'They represent great value for money . . .' said the rep, seizing a defence.

'Aye, 'cos nowt can eat 'em, they are rock-hard.'

Eventually, after a dousing with treacle, the heat of the summer, the passing of time and a few hungry horses having a go, the everlasting feed blocks did disappear, but it was a while before Santa delivered any more for the sheep.

Although outwardly there are no signs that Christmas Day is anything other than a normal one, there is still a special feeling in the air, as well as the smell of roasting turkey. We're proud to have done our chores. I've always told the children, 'We look after the sheep and they look after us.'

We come back inside later in the morning to relax and enjoy the frivolities and food, knowing that all is well outside.

For Reuben, the best Christmas present ever was when our worst mechanical nightmare became a reality. The quad bike is our workhorse, used and abused on a daily basis, travelling some of the roughest terrain imaginable in all weathers, often towing trailers overloaded with hay bales and cake. If it's out of action, everything grinds to a halt. In the severest of weather, when the bike can't travel, we use the original form of horse power, coupled of course with Shanks's pony, to get to the sheep; but it's time-consuming and back-breaking. Returning from my last heaf of sheep on Christmas morning, I began to feel a vibration coming from the bike, and not a good one.

'Can yer feel that?' I said to Miles and Violet, who were perched either side of me.

'What?' said Miles. Violet said nothing; she likely hadn't heard me from under her balaclava, over which she was wearing a knitted bobble hat. I didn't ask again, reckoning that they were preoccupied with what Santa might have left for them under the Christmas tree rather than whether the bike was making an unhealthy noise. I put it to the back of my mind and went back to the farmhouse with the children, ready for the start of present-opening. Clive had his last lot of sheep to feed, and sped off on the bike towards Birkdale Common.

'Won't be lang,' he shouted. 'Don't start without mi.'

The children were very keen for him to get back, so the present-opening could start. But it seemed like forever before we heard the quad bike coming back up the yard, very slowly. Looking through the tiny front-porch window I could see Clive was walking alongside the bike, leaning across to operate the throttle and steer it home. Even from afar I could tell from his body language that he wasn't in the best of humours, and as he got closer still I could see why: the front wheel was distinctly skew-whiff. Every so often Clive and the bike would stop; Clive would administer a kick to the wheel to line it up again, and then he'd move it a bit further.

I popped my head out of the front door and scowled at him.

'C'mon, Clive,' I complained. 'The kids 'ave waited lang enough to open their presents now.'

A string of angry words was aimed back at me. Apparently the track rod end had disconnected, meaning that there was only one wheel steering, the other going whichever way it chose. The children resigned themselves to waiting a bit longer while Clive got the bike back into the building – all of them bar one: Reuben. Pulling on his coat, hat and wellies, he was out of the door like a shot.

'I think I can mend it, Dad,' he said, his eyes lighting up at the prospect of lying on the straw in the barn under the quad bike on Christmas Day.

'I doubt thoo can, Reubs,' said Clive flatly. 'An' there's no way I can get any parts for it until after Christmas. It's a bad job.' And with that, he went huffily into the house.

As far as Reuben was concerned, the gauntlet had been thrown down: he *had* to repair it now. He did come back inside to open his presents (one of which was a tool box), but his mind was elsewhere, puzzling over how he could get the bike up and running again. He went back outside as soon as he could. While I was in the kitchen preparing the sprouts and potatoes for our evening meal I could hear him hammering away at something, but rather than worry about it I decided that he couldn't possibly make the situation any worse, so he might just as well get on with it. As long as he was happy. Clive wasn't fussed at all by now: cup of tea in hand, blazing fire, contented children, Queen's speech, an *Only Fools and Horses* repeat Christmas special on the TV. He had resigned himself to spending the next couple of days without the bike.

A few hours went by. It was almost time to head out and do the evening feeding rounds before Christmas dinner was served, when Reuben came flying in, beaming.

'It's mended,' he said. 'I've sorted the bike out.'

Clive was doubtful, but went along with it.

''Ave yer, mi owd mate,' he said, patting Reuben's head. Reuben was oblivious to Clive's scepticism. Out we all went, Reuben half skipping, unable to contain himself.

The bike's wheels looked straight. Clive checked underneath, rocked the broken wheel, then stood up, raised his eyebrows and started the bike. Setting off slowly, he steered to the left, then the right, then did a full circle back around into the barn.

'Reubs, I 'ave no idea 'ow you've done it, but well done,' he said.

Reuben set off on a very long, very detailed blow-by-blow explanation of his unorthodox repair, which involved him robbing bits from an old pushbike, using a tapered nut and washer, and the ingenious deployment of a G-clamp. His love of watching *Scrapheap Challenge* on winter nights after school had clearly paid dividends. It was obviously just a temporary repair, but Reuben was brimming with pride, and his achievement lasted until we could get it repaired properly.

We hope to wade through the turkey by New Year, but it can be a close-run thing. New Year's Eve tends to be a quiet affair at Ravenseat, with our toughest winter months still ahead of us. Invariably we spend the day outside in the cold and on coming into the warm house and settling in front of the fire, all intentions of seeing the New Year in fall by the wayside. One by one we carry the children off to their beds. For the last few years on New Year's Day, weather permitting, we've saddled up the horses and gone for a ride, sometimes calling on our neighbours to wish them well for the forthcoming year.

When all is said and done, we ourselves are only temporary custodians, passing through. The seasons change and the years fly by, and through working the land I feel a connection with those who went before. The traditions that link us intrinsically to the past remain unbroken. The privilege I feel to live and work here is clearly a sentiment that was shared by others over the centuries: there, scratched into a beam above a barn door, are the words: 'mine eyes unto the hills'.

They are the same hills and windswept moors that we see every day – and they're still grazed by the Swaledale sheep, descended from those original flocks. Neither Clive nor I are natives of these parts, but we have had the good fortune to stumble upon this place and each other. Now we, too, have put down our roots, hopefully ready for the next generation to take on the mantle of inhabitants and guardians of Ravenseat.

Acknowledgements

With thanks to Ingrid Connell, Jean Ritchie, Jo Cantello, Rachel Hall, Elenor Alderson, Jenny Harker and Colin and Anne Martin.

If you enjoyed
*A Year in the Life of the
Yorkshire Shepherdess*
look out for the new book by
bestselling author Amanda Owen

**Return to the farm with Yorkshire's favourite shepherdess as
she juggles the chaos of farming and family life.**

In *Adventures of the Yorkshire Shepherdess* Amanda takes us from
her family's desperate race to save a missing calf to finding her bra
has been repurposed as a house martin's nest, and from wild
swimming to the brutal winter of 2018 that almost brought her
to her knees. As busy as she is with her family and flock though,
an exciting new project soon catches her eye . . .

Ravenseat is a tenant farm and may not stay in the family, so
when Amanda discovers a nearby farmhouse up for sale, she
knows it is her chance to create roots for her children. The old
house needs a lot of renovation and money is tight, so Amanda
sets about the work herself, with some help from a travelling
monk, a visiting plumber and Clive. It's fair to say things do not
go according to plan!

Funny, evocative and set in a remote and beautiful landscape, this
new book will delight anyone who has hankered after a new life
in the country.